Common Diseases Of Farm Animals

by

R. A. Craig

Common Diseases Of Farm Animals
by R. A. Craig

ISBN: 978-93-59950-43-3

Published by

DOUBLE 9 BOOKS

2/13-B, Ansari Road
Daryaganj, New Delhi – 110002
info@double9books.com
www.double9books.com
Tel. 011-40042856

ABOUT THE AUTHOR

R. A. Craig, the author of "Common Diseases of Farm Animals," is widely known for his profound literary contributions that increase a long way past the world of veterinary science. With this seminal work, he showcases his versatility as a writer and philosopher, transcending the boundaries of mere technical literature. A genuine wordsmith, R. A. Craig possesses the top notch ability to weave ancient evaluation seamlessly into his writings. Through his eloquent prose and meticulous studies, he not most effective imparts important knowledge about farm animal illnesses however also invites readers on a adventure through time, connecting the beyond with the present. His willpower to bridging ancient contexts with contemporary issues enriches the reader's know-how and fosters a deeper reference to the difficulty rely. In the world of literature, R. A. Craig's works serve as a conduit for human connection and empathy. His creativity and passion shine thru his writing, presenting readers a glimpse into a myriad of emotions and reports. His narratives are a testomony to his storytelling prowess, portray brilliant landscapes of the human and animal situation. Above all, R. A. Craig's writings embody beauty and accessibility, making complex topics approachable to a huge audience.

CONTENTS

PREFACE

In preparing the material for this book, the author has endeavored to arrange and discuss the subject matter in a way to be of the greatest service and help to the agricultural student and stockman, and place at their disposal a text and reference book.

The general discussions at the beginning of the different sections and chapters, and the discussions of the different diseases are naturally brief. An effort has been made to conveniently arrange the topics for both practical and class-room work. The chapters have been grouped under the necessary heads, with review questions at the end of each chapter, and the book divided into seven parts.

The chapters on diseases of the locomotory organs, the teeth, surgical diseases and castration, although not commonly discussed in books of this class, the writer believes will be of value for reference and instructional work.

When used as a text-book, it will be well for the instructor to supplement the text with class-room discussions.

The writer has given special emphasis to the cause and prevention of disease, and not so much to the medicinal treatment. Stockmen are not expected to practise the medicinal treatment, but rather the preventive treatment of disease. For this reason it is not deemed advisable to give a large number of formulas for the preparation of medicinal mixtures to be used for the treatment of disease, but such treatment is suggested in the most necessary cases.

R. A. CRAIG.

PURDUE UNIVERSITY, LaFayette, Ind. August, 1915.

PART I
INTRODUCTORY

CHAPTER I
GENERAL DISCUSSION OF DISEASE

Disease is the general term for any deviation from the normal or healthy condition of the body. The morbid processes that result in either slight or marked modifications of the normal condition are recognized by the injurious changes in the structure or function of the organ, or group of body organs involved. The increase in the secretion of urine noticeable in horses in the late fall and winter is caused by the cool weather and the decrease in the perspiration. If, however, the increase in the quantity of urine secreted occurs independently of any normal cause and is accompanied by an unthrifty and weakened condition of the animal, it would then characterize disease. Tissues may undergo changes in order to adapt themselves to different environments, or as a means of protecting themselves against injuries. The coat of a horse becomes heavy and appears rough if the animal is exposed to severe cold. A rough, staring coat is very common in horses affected by disease. The outer layer of the skin becomes thickened when subject to pressure or friction from the harness. This change in structure is purely protective and normal. In disease the deviation from normal must be more permanent in character than it is in the examples mentioned above, and in some way prove injurious to the body functions.

CLASSIFICATION.—We may divide diseases into three classes: *non-specific, specific* and *parasitic.*

Non-specific diseases have no constant cause. A variety of causes may produce the same disease. For example, acute indigestion may be caused by a change of diet, watering the animal after feeding grain, by exhaustion and intestinal worms. Usually, but one of the animals in the stable or herd is affected. If several are affected, it is because all have been subject to the same condition, and not because the disease has spread from one animal to another.

Specific Diseases.—The terms infectious and contagious are used in speaking of specific diseases. Much confusion exists in the popular use of these terms. A *contagious* disease is one that may be transmitted by personal contact, as, for example, influenza, glanders and hog-cholera. As these diseases may be produced by indirect contact with the diseased animal as well as by direct, they are also *infectious*. There are a few germ diseases that are not spread by the healthy animals coming in direct contact with the diseased animal, as, for example, black leg and southern cattle fever. These are purely infectious diseases. Infection is a more comprehensive term than contagion, as it may be used in alluding to all germ diseases, while the use of the term contagion is rightly limited to such diseases as are produced principally through individual contact.

Parasitic diseases are very common among domestic animals. This class of disease is caused by insects and worms, as for example, lice, mites, ticks, flies, and round and flat worms that live at the expense of their hosts. They may invade any of the organs of the body, but most commonly inhabit the digestive tract and skin. Some of the parasitic insects, mosquitoes, flies and ticks, act as secondary hosts for certain animal microorganisms that they transmit to healthy individuals through the punctures or the bites that they are capable of producing in the skin.

CAUSES.—For convenience we may divide the causes of disease into the predisposing or indirect, and the exciting or direct.

The predisposing causes are such factors as tend to render the body more susceptible to disease or favor the presence of the exciting cause. For example, an animal that is narrow chested and lacking in the development of the vital organs lodged in the thoracic cavity, when exposed to the same condition as the other members of the herd, may contract disease while the animals having better conformation do not . Hogs confined in well-drained yards and pastures that are free from filth, and fed in pens and on feeding floors that are clean, do not become hosts for large numbers of parasites. Hogs confined in filthy pens are frequently so badly infested with lice and intestinal worms that their health and thriftiness are seriously interfered with. In the first case mentioned the predisposition to disease is in the individual, and in the second case it is in the surroundings .

The exciting causes are the immediate causes of the particular disease. Exciting causes usually operate through the environment. With the exception of the special disease-producing germs, the most common exciting causes are faulty food and faulty methods of feeding. The following predisposing causes of disease may be mentioned:

Age is an important factor in the production of disease. Young and immature animals are more prone to attacks of infectious diseases than are old and mature animals. Hog-cholera usually affects the young hogs in the herd first, while scours, suppurative joint disease and infectious sore mouth are diseases that occur during the first few days or few weeks of the animal's life. Lung and intestinal parasites are more commonly found in the young, growing animals. Old animals are prone to fractures of bones and degenerative changes of the body tissues. As a general rule, the young are more subject to acute diseases and the old to chronic diseases.

The surroundings or environments are important predisposing factors. A dark, crowded, poorly ventilated stable lowers the animal's vitality, and renders it more susceptible to the disease. A few rods difference in the location of stables and yards may make a marked difference in the health of the herd. A dry, protected site is always preferable to one in the open or on low, poorly drained soil. The majority of domestic animals need but little shelter, but they do need dry, comfortable quarters during wet, cold weather.

Faulty feed and faulty methods of feeding are very common causes of diseases of the digestive tract and the nervous system. A change from dry feed to a green, succulent ration is a common cause of acute indigestion in both horses and cattle. The feeding of a heavy ration of grain to horses that are accustomed to exercise, during enforced rest may cause liver and kidney disorders. The feeding of spoiled, decomposed feeds may cause serious nervous and intestinal disorders.

One attack of a certain disease may influence the development of subsequent attacks of the same, or a different disease. An individual may suffer from an attack of pneumonia that so weakens the disease-resisting powers of the lungs as to result in a tubercular infection of these organs. In the horse, one attack of azoturia predisposes it to a second attack. One attack of an infectious disease usually confers immunity against that particular disease. *Heredity* does not play as important a part in the development of diseases in domestic animals as in the human race. A certain family may inherit a predisposition to disease through the faulty or insufficient development of an organ or group of organs. The different species of animals are affected by diseases peculiar to that particular species. The horse is the only species that is affected with azoturia. Glanders affects solipeds, while black leg is a disease peculiar to cattle.

QUESTIONS

1. What is disease?

2. How are diseases classified? Give an example of the different classes.

3. What is a predisposing cause? Exciting cause?

4. Name the different predisposing and exciting causes of disease.

CHAPTER II
DIAGNOSIS AND SYMPTOMS OF DISEASE

The importance of recognizing or diagnosing the seat and nature of the morbid change occurring in an organ or group of organs cannot be overestimated. Laymen do not comprehend the difficulty or importance of correctly grouping the signs or symptoms of disease in such a way as to enable them to recognize the nature of the disease. In order to be able to understand the meaning of the many symptoms or signs of disease, we must possess knowledge of the structure and physiological functions of the different organs of the body. We must be familiar with the animal when it is in good health in order to be able to recognize any deviation from the normal due to disease, and we must learn from personal observation the different symptoms that characterize the different diseases. Stockmen should be able to tell when any of the animals in their care are sick as soon as the first symptom of disease manifests itself, by changes in the general appearance and behavior. But in order to ascertain the exact condition a general and systematic examination is necessary. The examiner, whether he be a layman or a veterinarian, must observe the animal carefully, noting the behavior, appearance, surroundings, and general and local symptoms.

Before making a *general examination* of the animal it is well, if the examiner is not already acquainted with the history of the case (care, feed and surroundings), to learn as much about this from the attendant as is possible. Inquiry should be made as to the feeding, the conditions under which the animal has been kept, the length of time it has been sick, its actions, or any other information that may be of assistance in forming the diagnosis and outlining the treatment.

The *general symptoms* inform us regarding the condition of the different groups of body organs. A careful study of this group of symptoms enables us correctly to diagnose disease and inform ourselves as to the progress of long, severe affections. These symptoms occur in connection with the pulse, respirations, body temperature, skin and coat, visible mucous membranes, secretions and excretions, and behavior of the animal.

The local symptoms are confined to a definite part or organ. Swelling, pain, tenderness and loss of function are common local symptoms. A *direct* symptom may also be considered under this head because of its direct relation to the seat of disease. It aids greatly in forming the diagnosis.

Other terms used in describing symptoms of disease are *objective*, which includes all that can be recognized by the person making the examination; *indirect*, which are observed at a distance from the seat of the disease; and *premonitory*, which precede the direct, or characteristic symptoms. The *subjective* symptoms include such as are felt and described by the patient. These symptoms are available from the human patient only.

Pulse.—The character of the intermittent expansion of the arteries, called the pulse, informs us as to the condition of the heart and blood-vessels. The frequency of the pulse beat varies in the different species of animals. The smaller the animal the more frequent the pulse. In young animals the number of beats per minute is greater than in adults. Excitement or fear, especially if the animal possesses a nervous temperament, increases the frequency of the pulse. During, and for a short time after, feeding and exercise, the pulse rate is higher than when the animal is standing at rest.

The following table gives the normal rate of the pulse beats per minute:

Horse 36 to 40 per minute
Ox 45 to 50 per minute
Sheep 70 to 80 per minute
Pig 70 to 80 per minute
Dog 90 to 100 per minute

In sickness the pulse is instantly responsive. It is of the greatest aid in diagnosing and in noting the progress of the disease. The following varieties of pulse may be mentioned: *frequent, infrequent, quick, slow, large, small, hard, soft* and *intermittent*. The terms frequent and infrequent refer to the number of pulse beats in a given time; quick and slow to the length of time required for the pulse wave to pass beneath the finger; large and small to the volume of the wave; hard and soft to its compressibility; and intermittent to the occasional missing of a beat. A pulse beat that is small and quick, or large and soft, is frequently met with in diseases of a serious character.

The horse's pulse is taken from the submaxillary artery at a point anterior to, or below the angle of the jaw and along its inferior border . It is here that the artery winds around the inferior border of the jaw in an upward direction, and, because of its location immediately beneath the skin, it can be readily located by pressing lightly over the region with the fingers.

Cattle's pulse is taken from the same artery as in the horse. The artery is most superficial a little above the border of the jaw. It is more difficult to find the pulse wave in cattle than it is in horses, because of the larger amount of connective tissue just beneath the skin and the heavier muscles of the jaw. A very satisfactory pulse may be found in the small arteries located along the inferior part of the lateral region of the tail and near its base.

The sheep's pulse may be taken directly from the femoral artery by placing the fingers over the inner region of the thigh. By pressing with the hand over the region of the heart we may determine its condition.

The hog's pulse can easily be taken from the femoral artery on the internal region of the thigh. The artery crosses this region obliquely and is quite superficial toward its anterior and lower portion.

The dog's pulse is usually taken from the brachial artery. The pulse wave can be readily felt by resting the fingers over the inner region of the arm and just above the elbow. The character of the heart beats in dogs may be determined by resting the hand on the chest wall.

RESPIRATION.—The frequency of the respirations varies with the species. The following table gives the frequency of the respirations in domestic animals:

Horse 8 to 10 per minute
Ox 12 to 15 per minute
Sheep 12 to 20 per minute
Dog 15 to 20 per minute
Pig 10 to 15 per minute

The ratio of the heart beats to the respirations is about 1:4 or 1:5. This ratio is not constant in ruminants. Rumination, muscular exertion and excitement increase the frequency and cause the respirations to become irregular. In disease the ratio between the heart beats and respirations is greatly disturbed, and the character of the respiratory sounds and movements may be greatly changed .

Severe exercise and diseased conditions of the lungs cause the animal to breathe rapidly and bring into use all of the respiratory muscles. Such forced or labored breathing is a common symptom in serious lung diseases, "bloat" in cattle, or any condition that may cause dyspnoea. Horses affected with "heaves" show a double contraction of the muscles in the region of the flank during expiration. In spasm of the diaphragm or "thumps" the expiration appears to be a short, jerking movement of the flank. In the abdominal form of respiration the movements of the walls of the chest are limited. This occurs in pleurisy. In the thoracic form of respiration the abdominal wall is held rigid and the movement of the chest walls make up for the deficiency. This latter condition occurs in peritonitis.

A cough is caused by irritation of the membrane lining the air passages. The character of the cough may vary according to the nature of the disease. We may speak of a moist cough when the secretions in the air passages are more or less abundant. A dry cough occurs when the lining membrane of the air passages is dry and inflamed. This may occur in the early stage of the inflammation, or as a result of irritation from dust or irritating gases. Chronic cough occurs when the disease is of long duration or chronic. In pleurisy the cough may be short and painful, and in broken wind, deep and suppressed. In parasitic diseases of the air passages and lungs, the paroxysm of coughing may be severe and "husky" in character.

The odor of the expired air, the character of the discharge and the respiratory sounds found on making a careful examination are important aids in arriving at a correct diagnosis, and in studying the progress of the disease.

Body Temperature.—The body temperature of an animal is taken by inserting the fever thermometer into the rectum. In large animals a five-inch, and in small animals a four-inch fever thermometer is used. It should be inserted full length and left in position from one and one-half to three minutes, depending on the rapidity with which it registers .

The average normal body temperatures of domestic animals are as follows:

    Horses 100.5\260 F.
    Cattle 101.4\260 F.
    Sheep 104.0\260 F.
    Swine 103.0\260 F.
    Dog 101.4\260 F.

There is a wide variation in the body temperatures of domestic animals. This is especially true of cattle, sheep and hogs. In order to determine the normal temperature of an animal, it may be necessary to take two or more readings at different times, and compare them with the body temperatures of other animals in the herd that are known to be healthy.

Exercise, feeding, rumination, excitement, warm, close stables, exposure to cold and drinking ice cold water are common causes of variations in the body temperatures of domestic animals.

Visible Mucous Membranes.—The visible mucous membranes, as they are termed, are the lining membranes of the eyelids, nostrils and nasal cavities, and mouth. In health they are usually a pale red, excepting when the animal is exercised or excited, when they appear a brighter red and somewhat vascular. In disease the following changes in color and appearance may be noted: When inflamed, as in cold in the head, a deep red; in impoverished or bloodless conditions of the body and in internal haemorrhage, pale; in diseases of the liver, sometimes yellowish, or dark red; in diseases of the digestive tract (buccal mucous membrane), coated; if inflamed, dry at first, later excessively moist; and in certain germ diseases a mottled red, or showing nodules, ulcers and scars.

Surface of the Body.—When a horse is in a good condition and well cared for, the coat is short, fine, glossy and smooth and the skin pliable and elastic. Healthy cattle have a smooth, glossy coat and the skin feels mellow and elastic. The fleece of sheep should appear smooth and have plenty of yolk, the skin pliable and light pink in color. When the coat loses its lustre and gloss and the skin becomes hard, rigid, thickened and dirty, it indicates a lack of nutrition and an unhealthy condition of the body. In sheep, during sickness, the wool may become dry and brittle and the skin pale and rigid. When affected with external parasites, the hair or wool becomes dirty and rough, a part of the skin may be denuded of hair, and it appears thickened, leathery and scabby, or shows pimples, vesicles and sores.

During fever, the temperature of the surface of the body is very unequal. In serious diseases or diseases that are about to terminate fatally, the skin feels cold and the hair is wet with sweat.

When animals are allowed to "rough it" during the cold weather, the coat of hair becomes heavy and rough. This is a provision of nature and enables them, as long as the coat is dry, to withstand severe cold.

Horses that are in a low physical condition, or when accustomed to hard work, if then kept in a stall for a few days without exercise, commonly show a filling of the cannon regions of the posterior extremities. This condition also commonly occurs in disease and in mares that have reached the latter period of pregnancy. Sheep that are unthrifty and in a poor physical condition, especially if this is due to internal parasites, frequently develop dropsical swellings in the region of the jaw, or neck.

Body Excretions.—The character of the body excretions, faeces and urine may become greatly changed in certain diseases. It is important that the stockman or veterinarian observe these changes, and in certain diseases make an analysis of the urine. This may be necessary in order properly to diagnose the case.

Behavior of the Animal.—When the body temperature is high, the animal may appear greatly depressed. If suffering severe pain, it may be restless. In diseases of the nervous system, the behavior of the animal may be greatly changed. Spasms, convulsions, general local paralysis, stupid condition and unconsciousness may occur as symptoms of this class of disease.

QUESTIONS

1. What information is necessary in order to be able to recognize or diagnose disease? 2. What are the general symptoms of disease?

3. What are the subjective symptoms of disease?

4. Describe method of taking the pulse beat in the different animals and its character in health and disease.

5. Give the ratio of the heart beats to the respirations in the different species of animals.

6. What are the normal body temperatures in the different domestic animals?

7. What are the visible mucous membranes?

8. Is the condition of the coat and skin any help in the recognition of disease?

CHAPTER III
TREATMENT

Preventive Treatment.—The subject of preventive medicine becomes more important as our knowledge of the cause of disease advances. A knowledge of feeds, methods of feeding, care, sanitation and the use of such biological products as bacterins, vaccines and protective serums is of the greatest importance to the farmer and veterinarian. We are beginning to realize that one of the most important secrets of profitable and successful stock raising is the prevention of disease; that the agricultural colleges are doing a great work in helping to teach farmers that there are right and wrong methods of feeding and caring for animals; that the practice of sanitation in caring for animals is the cheapest method of treating disease; and that it is advisable to practise radical methods of control, when necessary, in order to rid the herd of an infectious disease.

The ration fed and the method of feeding are not only important in considering the causes of diseases of the digestive tract, but diseases of other organs as well. The feeding of an excessive, or insufficient quantity of feed, or a ration that is too concentrated, bulky and innutritious, poor in quality, or spoiled may produce disease.

An impure water supply is a common cause of disease. A deep well that is closed in properly and does not permit of contamination from filth, does not insure a clean water supply if the trough or tank is not kept clean.

Farm Buildings.—If stockmen would make a more careful study of the kind of farm buildings most suitable to their needs, the selection of the location, the proportions, the arrangement of the interior and the lighting and ventilation, there would be a great saving in losses from disease, and the cost of building in many cases would be lessened. Your neighbor's building that you have taken for your model may not be suitable for your needs. It may be more expensive than your financial condition permits. It may be poorly lighted and ventilated and not suited to the site that you have selected.

Biological Products.—There are a number of biological products that may be used in the prevention and control of disease. Some of these products, such as tuberculin and malein, enable the owner to rid his herds of tubercular cows and glandered horses before these diseases have become far enough advanced to be recognized by the visible symptoms alone. Black leg, anthrax and hog-cholera vaccines are valuable agents in the control of disease. In the treatment of fistula and infectious abortion, bacterins may be used. There are many other germ diseases and infections for which vaccines and bacterins may be used. However, we must not depend wholly on these agents in the control of disease. We must possess a knowledge of the manner in which the infection is spread, for without this knowledge we would be unable to prevent its dissemination over a wide area.

Medicinal Treatment.—The average stockman or veterinarian is more familiar with the treatment of disease with drugs than he is with the preventive measures just described. This statement does not imply that a knowledge of medicinal therapeutics is not of the greatest importance in the treatment of disease. The ultimate object of all drugs is both to prevent and cure disease, but the injudicious use of a drug does neither. A discussion of this subject cannot be entered into here, and because of its largeness it is not advisable to discuss it further than a brief summary of the methods of administering drugs.

Administration of Drugs.—Drugs may be administered by the following channels: by way of the mouth, in the feed or as a drench; by injecting into the tissues beneath the skin or hypodermically; by rubbing into the skin; by the air passages and the lungs; and by injecting into the rectum.

If the animal is not too sick to eat and the drug does not possess an unpleasant taste, it may be given with the feed. If soluble, it may be given with the drinking water, or in any case, it may be mixed with ground feed if this method is to be preferred. In all cases the medicine must be well mixed with the feed. This is especially important if there are a number of animals to be treated, as there is more certainty of each animal getting the proper dose and the danger of overdosing is avoided. If the young animal is nursing the mother, we can take advantage of certain drugs being eliminated in the mother's milk and administer the drug to the mother.

DRENCHES.—In the larger animals a bulky drench is sometimes difficult to administer, and we should, in all cases, count on a portion being wasted.

Horses are sometimes difficult to drench, and it may be advisable to confine the horse in some way. Small drenches can readily be given with a syringe or a small bottle. In giving bulky drenches it is most convenient to use a long-necked, heavy glass bottle. The horse should be backed into a narrow stall and the head elevated by placing a loop in the end of a small rope over the upper jaw, passing the rope back of the nose piece on the halter and throwing it over a beam, and raising the head until the mouth is slightly higher than the throat. If the horse refuses to swallow, a tablespoonful of clean water may be dropped into the nostril. This forces it to swallow. A drench should never be given through the nose, as it may pass into the air passages and cause a fatal inflammation of the lungs.

Cattle can be easily drenched by taking hold of the nostrils with the fingers, or snapping a bull ring into the partition between the nostrils and elevating the head.

Sheep may be drenched either in the standing position, or when thrown on the haunches and held between the knees. Care should be exercised in giving irritating drenches to sheep, especially if the drench be bulky.

A herd of hogs may be quickly and easily drenched if they are confined in a small pen, and the loop of a small rope placed around the snout, well back toward the corners of the mouth. A small metal dose syringe should be used. If the drench is bulky and the hog difficult to hold, it may be necessary to elevate the head and raise the forefeet from the ground. The drench should not be given until the hog is quiet and well under control, as there is some danger of the medicine passing into the air passages and doing harm. It may be necessary to mark the hogs that have been drenched with a daub of paint, or in some other manner in order to be able to distinguish them from the untreated animals.

The administration of drugs enclosed in a gelatin capsule, or mixing them with syrup, honey or linseed oil, and rolling the mass into the form of a cylinder is commonly practised. The *capsule* or *ball* may then be shot into the pharynx with a balling gun. A ball may also be given to the larger animals by carrying it into the back part of the mouth with the hand, and placing it on the back part of the tongue. In the horse this method of administration requires some practice. The tongue must be pulled well forward, the head held up, and the tongue released as soon as the ball is placed on the tongue, so that it may pass back into the pharynx.

The administration of drugs by *injecting beneath the skin* is suitable when the drug is non-irritating and the dose is small. Drugs administered in this way act promptly and energetically. The alkaloid or active principle of the drug is commonly used. A fold of the skin is picked up with the fingers and the needle is quickly introduced, care being taken not to prick or scratch the muscular tissue, as this causes some pain and makes the animal restless. In order to avoid abscess formation at the point of injection, the skin should be cleansed with a disinfectant and the syringe and needle sterilized before using.

Drugs are not absorbed through the unbroken skin, but when applied with friction, or when the outer layer is removed by blistering, absorption may take place. Liniments, blisters and *poultices* are the preparations used.

Volatile drugs, such as chloroform and ether, are absorbed quickly by the enormous vascular surface of the lungs. This class of drugs is administered for the purpose of producing general anaesthesia. *Anaesthetics* are indispensable in many surgical operations.

The administration of a drug in the form of *medicated steam* is quite useful in combating some respiratory diseases. In steaming large animals a pail about half full of boiling-hot water to which has been added about an ounce of coal-tar disinfectant, or whatever drug is required, is held within about one foot of the animal's nostrils. It is usually advisable to throw a light cover over the head and pail in order to direct the steam toward the nostrils. Dogs can be placed on a cane-seated chair and a pail or pan of boiling-hot water placed under it, and a sheet thrown over all.

Drugs are administered by way of the rectum when the animal can not be drenched, or the drug can not be given in any other way and when a local action is desired. An *enema* or *clyster* is a fluid injection into the rectum and is employed for the following purposes: to accelerate the action of a purgative; to stimulate the peristaltic movement of the intestines; to kill intestinal parasites; to reduce body temperature; to administer medicine; and to supply the animal with food. An enema may be administered by allowing water to gravitate into the rectum from a height of two or three feet or by using an injection pump. In the larger animals several feet of heavy walled rubber tubing carrying a straight nozzle at one end should be used. In administering an enema, the rectum should be emptied out with the hand and the nozzle of the syringe carried as far forward as possible. The operator should be careful not to irritate or tear the wall of the rectum.

Size of the Dose.—The doses recommended in the treatment of the different diseases, unless otherwise stated, are for mature animals. The dose for a colt one year of age is about one-third the quantity given the adult, two years of age one-half, and three years of age two-thirds. In well-matured colts a larger dose may be given. In cattle, the doses recommended are about the same. In the smaller animals the size of the dose may be based on the development and age of the animal. When the drug is administered at short intervals or repeated, the size of the dose should be reduced. The physiological action of some drugs may be changed by varying the size of the dose.

QUESTIONS

1. Give a general description of preventive treatment.

2. By what channels may drugs be administered?

3. How are drenches administered?

4. How are solid drugs administered?

5. What kind of drugs are administered hypodermically?

6. What is an enema?

7. What proportion of the dose of a drug recommended for the adult may be given to immature animals?

PART II
NON-SPECIFIC OR
GENERAL DISEASES

CHAPTER IV
DISEASES OF THE DIGESTIVE SYSTEM

The organs that form the digestive tract are the mouth, pharynx, oesophagus, stomach, intestines and the annexed glands, viz.: the salivary, liver, and pancreas. The development of these organs differs in the different species of animals. For example, solipeds possess a small, simple stomach and capacious, complicated intestines. Just the opposite is true of ruminants. The different species of ruminants possess a large, complicated stomach, and comparatively simple intestines. In swine we meet with a more highly developed stomach than that of solipeds and a more simple intestinal tract. Of all domestic animals the most simple digestive tract occurs in the dog. These variations in the development of the different organs of digestion, together with the difference in the character of the feed and method of feeding, cause a variation in the kind of diseases met with in the different species. The complicated stomach of ruminants predispose them to diseases of this portion of the digestive tract. Because of their complicated intestinal tract solipeds are prone to intestinal disease.

DISEASES OF THE MOUTH

GENERAL DISCUSSION.—The mouth is the first division of the digestive tract. It is formed by the lips, cheeks, palate, soft palate, tongue and teeth. Here the feed is acted on mechanically. It is broken up by the teeth and moved about until mixed with the saliva and put into condition to pass through the pharynx and along the oesophagus to the stomach. The mechanical change that the feed is subject to is very imperfect in dogs. In the horse it is a slow, thorough process, although greedy feeders are not uncommon. The first mastication in the ox is three times quicker than in horses, but the process of rumination is slow and thorough.

STOMATITIS.—Simple inflammation of the mouth is frequently met with in horses. Ulcerative or infectious inflammation commonly occurs in young, and occasionally in old, debilitated animals. This form of sore mouth will be discussed along with other infectious diseases, and the following discussion will be confined to the non-infectious form of the disease.

The causes are irritation from the bit, sharp teeth, irritating drenches, roughage that contains beards or awns of grasses and grains, and burrs that wound the lining membrane of the mouth. Febrile, or digestive disorders, or any condition that may interfere with feeding, may cause this disorder. In the latter cases the mucous membrane of the mouth is not cleansed by the saliva. Particles of feed may decompose and irritating organisms set up an inflammation. Putrid or decomposed slops, hot feeds, irritating drenches and drinking from filthy wallows are common causes of inflammation of the mouth in hogs.

The symptoms vary in the different cases and species. Slight or localized inflammation of the mouth is usually overlooked by the attendant. Lampas of horses may be considered a local inflammation involving the palate. Lacerations of the cheek or tongue by the teeth, or irritating feed, usually result in a slight interference with prehension and mastication and more or less salivation. Salivation from this cause should not be confused with salivation resulting from feeding on white clover.

In generalized inflammation of the mucous membrane, the first symptom usually noticed is the inability to eat. On examining the mouth we find the mucous membrane inflamed, hot and dry. A part may appear coated. In a short time the odor from the mouth is fetid. Following this dry stage of the inflammation is the period of salivation. Saliva dribbles from the mouth, and in severe cases it is mixed with white, stringy shreds of epithelium and tinged with blood. In less acute forms of the disease, we may notice little blisters or vesicles scattered over the lining membrane of the lips, cheeks and tongue.

The acute form of stomatitis runs a short course, usually a few days, and responds readily to treatment. Localized inflammation caused by irritation from teeth, or feeding irritating feeds, does not respond so readily to treatment.

The treatment is largely preventive and consists largely in removing the cause. When the mouth is inflamed, roughage should be fed rather sparingly, and soft feeds such as slops, mashes, or gruels given in place of the regular diet. Plenty of clean drinking water should be provided. In the way of medicinal treatment antiseptic and astringent washes are indicated. A four per cent water solution of boric acid may be used, or a one-half per cent water solution of a high grade coal-tar disinfectant. The mouth should be thoroughly irrigated twice daily until the mucous surfaces appear normal.

DEPRAVED APPETITE

A depraved appetite is met with in all species of farm animals, but it is especially common in ruminants. It should not be classed as a disease, but more correctly as a bad habit, or symptom of innutrition or indigestion. The animals affected seem to have an irresistible desire to lick, chew and swallow indigestible and disgusting objects.

The common cause of depraved appetite is the feeding of a ration deficient in certain food elements. A ration deficient in protein or in salts is said to cause this disorder. Lack of exercise, or confinement, innutrition, and a depraved sense of taste may favor the development of this disease. For example, when sheep are housed closely they may contract the habit of chewing one another's fleeces. Lambs are especially apt to contract this habit when suckling ewes that have on their udders long wool soiled with urine and faeces.

The first symptom is the desire to chew, lick or eat indigestible or filthy substances. Horses and cattle may stand and lick a board for an hour or more; cattle may chew the long hair from the tails of horses; sheep may nibble wool; sows may within a short time after giving birth to their pigs, kill and eat them; chickens may pick and eat feathers. Innutrition may accompany the abnormal appetite, as very frequently the affected animal shows a disposition to leave its feed in order to eat these injurious and innutritions substances. In ruminants, the wool or hair may form balls and obstruct the opening into the third compartment, causing chronic indigestion and death.

The treatment consists in the removal of the cause. Feeding a ration that meets the needs of the system, clean quarters and plenty of exercise are the most important preventive lines of treatment. In such cases medicinal treatment (saline and bitter tonics) may be indicated. It is usually advisable to remove the affected animals from the herd or flock in order to prevent others from imitating them.

DISEASES OF THE STOMACH

There is a remarkable difference in the development of the stomachs of solipeds and ruminants.

The horse's stomach is simple and has a capacity of three or four gallons. The left portion is lined with a cuticular mucous membrane, and the right portion with a glandular mucous membrane that has in it the glands that secrete the gastric juice. The most important digestive change in the feed is the action of the gastric juice on the proteids and their conversion into the simpler products, proteoses and peptones.

RUMINANTS have a compound stomach (Figs. 9 and 10). The capacity of the stomach of the ox is between twenty and thirty gallons. The four compartments into which it is divided are the rumen, reticulum, omasum, and abomasum or true stomach. The rumen is the largest compartment, with a capacity of more than twenty gallons. The reticulum is the smallest, with a capacity of about one-half gallon.

After a brief mastication, the food passes directly to the *rumen*. Here it is subjected to a churning movement that mixes and presses the contents of the rumen forward in the direction of the oesophageal opening, where it is ready for regurgitation. It is then carried back to the mouth, remasticated and returned to the rumen. This is termed rumination. All food material that is sufficiently broken up is directed toward the opening into the third compartment by the oesophageal grove , a demi-canal that connects this with the oesophageal opening.

The third compartment, the *omasum*, communicates anteriorly with the second and first, and posteriorly with the fourth compartment or true stomach. The interior arrangement of this compartment is most singular. It is divided by a number of large folds of the lining membrane between which are smaller folds. It is between these folds that the contents pass.

The first three compartments possess no glands capable of secreting a digestive juice. However, important digestive changes occur. The carbohydrates are digested by means of enzymes contained in the feed. The most important function of the rumen and omasum is the maceration of the fibrous substances, and the digestion of the cellulose. Between sixty and seventy per cent of the cellulose is digested in the rumen.

The abomasum is lined by a gastric mucous membrane. The gastric juice secreted converts the protein into peptones. In the young a milk curdling ferment is also secreted by the glands of this compartment.

THE STOMACH OF THE HOG is a type between the carnivora and ruminant. The digestive changes may be divided into four stages. The first period is one of starch conversion; the second period is the same, only more pronounced; the third period, both starch and protein conversion occurs; and the fourth period is taken up mostly with protein digestion.

ACUTE INDIGESTION OF THE STOMACH OF SOLIPEDS.—Diseases of the stomach are less common in solipeds than in ruminants. The simple stomach of the horse and the comparatively unimportant place that it occupies in the digestion of the feed renders it less subject to disease. Only under the most unfavorable conditions for digestion of the feed does this class of disorders occur. Acute indigestion in the form of overloading and fermentation occurs in the stomach .

The predisposing causes that have to do with the development of these disorders, are the small capacity of the stomach and the location and smallness of the openings leading from the oesophagus and into the small intestines. Greedy eaters are more prone to indigestion than animals that eat slowly and are fed intelligently.

The following exciting causes may be mentioned: Sudden changes in ration; feeding too much green feed or grain; feeding frozen or decomposed feeds; drinking ice-cold water; and violent exercise or work that the animal is not accustomed to, immediately after feeding are the common disease-producing factors.

The symptoms may vary from impaired appetite and slight restlessness to violent, colicky pains. In the large majority of cases the attendant is unable to differentiate between this and other forms of acute indigestion. The characteristic symptoms are attempts at regurgitation and vomiting, assuming a dog-sitting position and finally such nervous symptoms as champing of the jaws, staggering movement and extreme dulness.

The violent form of gastric indigestion frequently ends in death. Rupture of the stomach is not an uncommon complication .

The treatment is both preventive and medicinal. This digestive disorder can be prevented. The feeding of the right kind of a ration and in the right way, and avoiding conditions that may interfere with the digestion of the feed, are the general lines of preventive treatment indicated. Such measures are of special importance in the handling of animals that possess an individual predisposition toward this class of disease. In mild attacks the animal should be subjected to a rigid or careful diet during the attack and for a few days later.

It is advisable to place the animal in a comfortable stall that is well bedded with straw and plenty large for it to move about in. If a roomy sick-stall can not be provided, a grass lot or barn floor may be used. If the weather is chilly or cold, the body should be covered with a blanket and roller bandages applied to the limbs.

Bulky drenches should not be given. Stimulants and drugs capable of retarding fermentation are indicated. Sometimes the administration of a sedative is indicated. Treatment should be prompt, as in many cases fermentation of the contents of the stomach occurs and gases form rapidly.

From two to four ounces of oil of turpentine may be given in from six to eight ounces of linseed oil.

ACUTE INDIGESTION OF THE STOMACH OF RUMINANTS.—The different forms of acute indigestion are bloating, overloading of the rumen and impaction of the omasum.

TYMPANITES, "BLOATING."—This disorder is usually caused by animals feeding on green feeds, such as clover, alfalfa and green corn, that ferment readily. Stormy, rainy weather seems to favor bloating. The consumption of spoiled feeds such as potatoes and beets may cause it. The drinking of a large quantity of water, especially if cold, chills the wall of the rumen and interferes with its movement. Frozen feeds may act in the same way. Sudden changes in the feed, inflammation of the rumen, and a weak peristaltic movement of the paunch resulting from disease or insufficient nourishment are frequent causes. It may occur in chronic disease. In tuberculosis, bloating sometimes occurs.

The symptoms are as follows: The paunch or rumen occupies the left side of the abdominal cavity, hence the distention of the abdominal wall by the collecting of gas in the rumen occurs principally on that side. The gas forms quickly and the distended wall is highly elastic and resonant. The animal stops eating and ruminating, the back may be arched and the ears droop. In the more severe cases the wall of the abdomen is distended on both sides, the respirations are quickened and labored, the pulse small and quick, the eyes are prominent and the mucous membrane congested. Death results from asphyxia brought on by the distended paunch pushing forward and interfering with the movement of the lungs and the absorption of the poisonous gases.

The treatment is both preventive and medicinal. This form of acute indigestion can be largely prevented by practising the following preventive measures: All changes in the feed should be made gradually, especially if the ration fed is heavy, or the new ration consists largely of green, succulent feed. Cattle pasturing on clover should be kept under close observation. It is not advisable to pasture cattle on rank growths of clover that are wet with dew or a light rain. Bloating can be quickly relieved by puncturing the wall of the paunch with the trocar and cannula. The operation is quite simple and is not followed by bad results. The instrument is plunged through the walls of the abdomen and rumen in the most prominent portion of the flank, midway between the border of the last rib and the point of the haunch . The trocar is then withdrawn from the cannula. After the gas has escaped through the cannula, the trocar is replaced and the instrument withdrawn. After using the trocar and cannula, the instrument should be cleaned by placing it in boiling hot water. It is advisable to wash the skin at the seat of the operation with a disinfectant before operating. In chronic tympanitis, it is sometimes advisable to leave the cannula in position by tying a tape to the flange, passing it around the body and tying.

As a cathartic for cattle, we may give one quart of linseed and from two to four ounces of turpentine, or one to two pounds of Epsom or Glauber's salts, dissolved in plenty of water. Sheep may be given about one-fourth the dose recommended for cattle.

OVERLOADING THE RUMEN. — This form of indigestion occurs when ruminants have access to feeds that they are not accustomed to. As a result, they eat greedily and the mass of feed in the rumen becomes so heavy that the walls of the organ can not move it about, and digestion is interfered with. This is especially true of succulent feeds. A diseased condition of the animal predisposes it to this disorder. If after eating an excessive amount of dry, innutritions fodder, the animal drinks freely of cold water, acute symptoms of overloading are manifested.

The general symptoms occurring in overloading resembles those seen in bloating. The symptoms may be mild and extend over a period of several days, or it may take on a highly acute form, terminating fatally within a few hours. The acuteness of the attack depends on the character and quantity of feed eaten. If a large quantity of green feed is eaten, fermentation occurs and the animal may die within a few hours. The swelling on the left side has a doughy feel. It is not as elastic and resonant as in bloat, even when complicated by some gas formation. The animal may stop ruminating, refuse to eat, and act dull. In the more severe cases the respirations are hurried and labored, the pulse small and quick and the expression of the face indicates pain. Colicky pains sometimes occur. Death may occur from shock or asphyxia.

The treatment is both preventive and curative. This disease can be prevented by using the necessary precautions to prevent animals from overeating. If gas forms, the trocar and cannula should be used. A drench of from one to two pounds of Epsom or Glauber's salts should be given. Sheep may be given from four to six ounces of Epsom or Glauber's salts. We should endeavor to stimulate the movement of the paunch by pressure on the flank with the hand, throwing cold water on the wall of the abdomen and by hypodermic injections of strychnine. Rumenotomy should be performed when necessary. This operation consists in opening the walls of the abdomen and rumen, and removing a part of the contents of the rumen. This is not a dangerous operation when properly performed, and should not be postponed until the animal is too weak to make a recovery.

IMPACTION OF THE OMASUM.—This disease may occur as a complication of other forms of acute indigestion and diseases accompanied by an abnormal body temperature. Feeds that are dry and innutritions commonly cause it. Other causes are an excessive quantity of feed, sudden changes in the diet and drinking an insufficient quantity of water.

As in other diseases of the stomach, the appetite is diminished, rumination ceases or occurs at irregular intervals, and the animal is more or less feverish. Bloating and constipation may occur. The animal may lose flesh, is weak, walks stiffly and grunts as though in pain when it moves about in the stall and at each respiration. In the acute form, marked symptoms are sometimes manifested. At first the animal acts drowsy; later violent nervous symptoms may develop.

The course of this disease varies from a few days to several weeks. Death frequently occurs. Frequently a diarrhoea accompanies recovery, a portion of the faeces appearing black with polished surfaces, as though they had been baked.

The preventive treatment consists in practising the necessary precautions against the development of this disease by avoiding sudden changes in the feed, the feeding of dry, innutritions feeds in too large amounts, allowing animals plenty of water and providing them with salt. The best purgative to give is Glauber's or Epsom salts in from one- to two-pound doses, dissolved in at least one gallon of water. This physic may be repeated in from twelve to eighteen hours if necessary. Two drachms of tincture of nux vomica and one ounce of alcohol may be given in a drench three times daily. Hypodermic injections of strychnine, eserine, or pilocarpine are useful in the treatment of this disease. When recovery begins, the animal should be allowed moderate exercise and be fed food of a laxative nature.

FOREIGN BODIES IN THE STOMACH OF RUMINANTS.—Foreign bodies such as hair balls and wire are very commonly found in the reticulum. This is because of the habits of this class of animals. Cattle eat their feed hastily and do not pick it over as carefully as does the horse.

Smooth, round objects do no appreciable harm unless they block the opening into the third compartment of the stomach. This frequently occurs in wool-eating lambs. Sharp-pointed objects may penetrate the surrounding tissues or such organs as the spleen, diaphragm, and pericardial sack. If these organs are injured by the foreign body serious symptoms develop. The *general symptoms* are pain, fever, weakness and marked emaciation. It is very difficult to form a correct diagnosis, as the disease comes on without any apparent cause. Sometimes a swelling is noticed in the right and inferior abdominal region. If the heart becomes injured, symptoms of pericarditis are manifested.

The treatment is largely preventive. Special care should be used to avoid getting foreign substances into the feed given to cattle. The feed troughs should be kept clean; we should avoid dropping nails and staples into the feed when repairing the silo or grain bin; and pieces of baling wire should be removed from straw or hay. Feeds known to be dirty should be run through a fanning mill before feeding.

INFLAMMATION OF THE STOMACH OF SWINE.—Overloading and feeding spoiled feed are *common causes* of inflammation of the stomach. Swill-fed hogs are most commonly affected with this disorder. Overloading more often results in an inflammation of the stomach if the overloading follows the feeding of a light ration, and the weather is extremely warm. Hogs that are accustomed to eating salt may eat too much of it when fed to them after it is withheld for a week or longer, and a large quantity of water is taken soon afterwards. Slop containing alkaline washing powders and soaps irritate the stomach and intestines and cause a serious inflammation.

The symptoms are loss of appetite, restlessness and sometimes colicky pains. The hog usually wanders off by itself, acts dull, grunts, lies down in a quiet place or stands with the back arched and the abdomen held tense. Vomiting commonly occurs. Sometimes the animal has a diarrhoea. The body temperature may be above normal.

The treatment consists in avoiding irritating feeds and sudden changes in the kind or quantity of feed fed. Drenching with hot water, or with about one ounce of ipecacuan may be practised. From one to three ounces of castor oil, depending on the size of the hog, may be given. After recovery the hogs should be confined in a comfortable pen and fed an easily digested ration.

DISEASES OF THE INTESTINES

GENERAL DISCUSSION.—The intestinal tract of solipeds is the best developed of any of the domestic animals . It is divided into two portions, *small* and *large*. The *small intestine* is a little over seventy feet in length and about one and one-half inches in diameter. The mucous membrane lining presents a large absorbing surface and is well supplied with absorbing vessels that take up the sugars, proteids and fats, which are finally distributed to the body cells by the blood capillaries. In addition to these absorbing vessels the mucous membrane contains intestinal glands that secrete the intestinal juice. Other digestive secretions from the pancreatic gland and the liver are poured into the small intestine near its origin. These digestive juices act on the proteids, sugars, starches and fats, changing them into substances that are capable of being absorbed.

After disengaging itself from the mass of loops lodged in the region of the left flank, the small intestine crosses to the region of the right flank, where it terminates in the first division of the large intestine.

The large intestine is formed by the following divisions: caecum, double colon, floating colon and rectum. The caecum is a large blind pouch that has a capacity of about seven gallons. The double colon is the largest division of the intestines. It is about twelve feet in length and has a capacity of about eighteen gallons. This portion of the intestine terminates in the region of the left flank in the floating colon. The latter is about ten feet in length and about twice the diameter of the small intestine, from which it can readily be distinguished by its sacculated walls. The rectum is the terminal portion of the intestinal tract. It is about one and one-half feet in length and possesses heavy, elastic walls.

Fermentation and cellulose digestion occur in the caecum and double colon. It is in the floating colon that the faeces are moulded into balls. The faeces are retained in the rectum until defecation takes place.

The *intestinal tract of cattle* is longer than that of solipeds and the different divisions are not as well defined as in the horse's intestine and about one-half its diameter. The large intestine is about thirty-five feet in length and its capacity six or seven gallons .

ACUTE INTESTINAL INDIGESTION OF SOLIPEDS.—Acute indigestion is more common in horses and mules than it is in any of the other domestic animals. Because of the difference in the causes and symptoms manifested, we may divide it into the following forms: spasmodic, flatulent and obstruction colic.

The predisposing causes are general and digestive debility resulting from the feeding of an insufficient or unsuitable ration, and general and parasitic diseases of the intestine. Nervous, well-bred horses are most susceptible to nervous or spasmodic colic.

The direct causes are improper methods of feeding and watering; giving the animal severe or unusual exercise immediately before or after feeding; the feeding of spoiled or green feeds and new grains; chilling of the body; imperfect mastication of feed because of defective teeth; obstruction of the intestine by worms.

The feeding of grain at a time when the animal is not in fit condition to digest it results in imperfect digestion in both the stomach and intestine. This leads to irritation of the intestine and abnormal fermentation of its contents.

The drinking of a large quantity of water immediately after feeding grain flushes at least a part of the undigested grain from the stomach through the small intestine and into the caecum. New grains, such as new oats, are hurried along the small intestine and reach the large intestine practically undigested. The two latter conditions are common causes of *flatulent* or *wind colic*. Sudden change in the ration, especially to a green feed, may result in intestinal irritation and flatulence.

Horses that are greedy feeders and have sharp, uneven, smooth or diseased teeth are unable to masticate the feed properly. This results in unthriftiness caused by imperfect digestion and assimilation of the feed. Such animals usually suffer from a catarrhal or chronic inflammation of the intestine, and may have periodic attacks of acute indigestion or colic.

Obstruction colic is very often caused by the feeding of too much roughage in the form of straw, shredded fodder, or hay. Debility often contributes to this form of indigestion, and the double colon may become badly impacted with alimentary matter.

Worms may irritate the intestinal mucous membrane and interfere with digestion, obstruct the intestine and cause debility and circulatory disturbances. The large round worm may form a tangled mass and completely fill a portion of the double colon.

Some species attach themselves to the intestinal wall, suck the blood of the host and cause anaemia and debility. Colic resulting from *circulatory disturbances* is not common. The female of a certain species of *strongulus* deposits eggs in the mucous membrane. On hatching, the larvae may enter a blood capillary, drift along in the blood stream and finally come to rest in a large blood-vessel that supplies a certain portion of the intestines with blood. Here the parasite develops. The wall of the vessel becomes irritated and inflamed, pieces of fibrin flake off and drift along the blood stream until finally a vessel too small for the floating particle to pass through is reached and the vessel becomes plugged. The loop of intestine supplied by it receives no blood. A temporary paralysis of the loop occurs, which persists until a second vessel is able to take over the function of the one that is plugged. This form of colic is most common in old horses .

Such complications of acute indigestion as *twisting, infolding* and *displacement of the intestine* may occur. It is not uncommon for a stallion to suffer from strangulated hernia, due to a rather large internal inguinal ring and a loop of the intestine passing through it and into the inguinal canal or scrotum. Such displacements are usually accompanied by severe colicky pains.

The symptoms vary in the different cases. In the mild form, the colicky pains are not prominent, but in the acute form, the animal is restless, getting up and down in the stall and rolling over. These movements are especially marked when the abdominal pain is severe.

In the spasmodic form the attack comes on suddenly, the colicky pains are severe, and the peristaltic movement of the intestine is marked and accompanied by loud intestinal sounds. In most cases of indigestion characterized by fermentation and collections of gas in the intestine there is gastric tympany as well.

Acute indigestion characterized by *impaction* of the large intestine pursues a longer course than the forms just mentioned, and the abdominal pain is not severe.

Congestion and inflammation of the intestine may result from the irritation produced by the feed. When this occurs, the abdominal pain is less violent. The animal usually acts dull, the walk is slow and unsteady, and the respirations and pulse beats may be quickened.

A large percentage of the cases of acute indigestion terminate fatally. The course of the disease varies from a few hours to several days.

The treatment is both *preventive* and *curative*. The preventive treatment is by far the most important. This consists in observing right methods of feeding and caring for horses. The attendant should note the condition of the animal before feeding grain, feed regularly and avoid sudden changes in feed. If a horse has received unusual exercise, it is proper to feed hay first, and when the animal is cooled out, water and feed grain. Drinking a small quantity of water when tired or following a meal is not injurious, but a large quantity of water taken at such times is injurious and dangerous to the health of the animal. The feeding of spoiled or mouldy feeds to horses is highly injurious.

The horse should be given a roomy, comfortable stall that is well bedded, or a clean grass lot. If the attack appears when the animal is in harness, we should stop working it and remove the harness immediately. Work or exercise usually aggravates the case and may cause congestion and inflammation of important body organs. In cold weather the animal should be protected by blankets. If the pain is violent, sedatives may be given. The gaseous disturbances should be relieved by puncturing the wall of the intestine with the trocar and cannula. Rectal injections of cold water may be resorted to. Fluid extract of cannabis indica in quarter ounce doses and repeated in one hour may be given in linseed oil. In all cases it is advisable to drench the animal with one pint of raw linseed oil and two ounces of turpentine. Strychnine, eserine and pilocarpine are the drugs commonly used by the veterinarians in the treatment of acute indigestion. Small and repeated doses of the above drugs are preferred to large doses. This is one of the diseases that requires prompt and skilled attention.

Sharp, uneven or diseased teeth should receive the necessary attention. In old horses, chopped hay or ground feeds should be fed when necessary. Debility resulting from hard work, wrong methods of feeding and intestinal disorders must be corrected before the periodic attacks of indigestion can be relieved. If the presence of intestinal worms is suspected, the necessary treatment for ridding the animal of these parasites should be resorted to.

Bitter or saline tonics should be administered in the feed when necessary. The following formula is useful as a digestive tonic: Sodium bicarbonate and sodium sulfate, one pound of each, powdered gentian one-half pound, and oil meal five pounds. A small handful of this mixture may be given with the feed two or three times daily.

INFLAMMATION OF THE INTESTINES.—The same causes mentioned in inflammation of the stomach and acute indigestion may cause this disease. It is most frequent at times when there are great variations in the temperature. Sudden cold or any influence that chills the surface of the body, or internal cold caused by drinking ice water or eating frozen feed, may cause it. The infectious forms of enteritis are caused by germs and ptomaines in the feed. Drinking filthy water or eating spoiled, mouldy feeds are common causes. In cattle pasturing in low, marshy places, enteritis may be common. The toxic form is caused by irritating poisons, such as caustic acids, alkalies and meat brine.

In the mild form of enteritis the appetite is irregular, the animal acts dull and stupid and may be noticed lying down more than common. Slight abdominal pains occur, especially following a meal. An elevation in the body temperature may be noted and the animal may drink more water than usual. Constipation or a slight diarrhoea may be present. The feces may be soft and foul smelling, coated with mucus, and slightly discolored with blood.

In the severe form of enteritis pressure on the abdomen may cause pain, the respiration and pulse beats are quickened and the body temperature is elevated. The abdominal pain may be severe and the animal is greatly depressed or acts dull. The movement of the intestines is suppressed at first and constipation occurs. Fermentation and the formation of gas may take place. Later the intestinal peristalsis increases and a foul-smelling diarrhoea sets in that is often mixed with blood. In the toxic form there may be marked nervous symptoms. Spasms, convulsions, stupefaction and coma may be manifested.

In the mild form recovery usually occurs within a few days. The more serious forms of the disease do not terminate so favorably. In the toxic form death usually occurs within a few days.

The large majority of cases of enteritis can be prevented by practising the necessary *preventive measures*. It is very necessary that animals exposed to cold be provided with dry sleeping quarters that are free from draughts. Where a number of animals are fed a heavy grain ration, or fed from the same trough, they should be kept under close observation. This is necessary in order to detect cases of indigestion or overfeeding early, and resort to the necessary lines of treatment, so as to prevent further irritation to the intestinal tract. Live stock should not be forced to drink water that is ice-cold. Low, poorly-drained land is not a safe pasture for cattle and horses. Spoiled roots, grains and silage, mouldy, dirty roughage and decomposed slops should not be fed to live stock.

The treatment consists in withholding all feed and giving the animal comfortable, quiet quarters—warm quarters and protection from the cold, providing the animal with a heavy straw bed, or with blankets if necessary, if the weather is cold. From five to forty grains of calomel may be given, depending upon the size of the animal and the frequency of the dose, two or three times a day. In case the animal is suffering severe pain, morphine given hypodermically may be indicated. In the mild form and at the very beginning of the attack, linseed oil may be administered to the larger animals. The dose is about one quart. The smaller animals may be given castor oil in from one- to four-ounce doses.

When convalescence is reached the animal should be fed very carefully, as the digestive tract is not in condition to digest heavy rations or feeds that ferment readily.

DIARRHOEA.—Diarrhoea occurs as a symptom of irritation and inflammation of the intestinal mucous membrane. Sudden changes in the feed, the feeding of a succulent green ration, severe exercise when the animal is not in condition for it, and chronic indigestion may cause diarrhoea in the absence of an intestinal inflammation.

The following symptoms may be noted: Animals affected by a diarrhoea act dull and weak; thirst is increased and the animal may show evidence of fever; the intestinal evacuations are soft, thin, and sometimes have an offensive odor. If the diarrhoea continues for several days, the animal loses flesh rapidly and the appetite is irregular. In such cases weakness is a prominent symptom.

Recovery usually occurs when the animal is dieted and rested.

The treatment consists in giving a physic of linseed or castor oil. Horses and cattle may be given from one-half to one quart of linseed oil; sheep and hogs from one to four ounces of castor oil. Feed should be withheld. Morphine may be given hypodermically to the large animals after a period of six to eight hours following the administration of the physic.

The following formula is quite useful in checking diarrhoea: salol one-half ounce, bismuth subnitrate one ounce, and bicarbonate of soda two ounces. The dose of this mixture is from one to four drachms, depending on the size of the animal, three or four times a day.

WHITE SCOURS OR DIARRHOEA IN YOUNG ANIMALS.—Young animals, when nursing the mother or fed by hand, frequently develop congestion and inflammation of the stomach and intestines. This disorder is characterized by a diarrhoea.

The causes may be grouped under two heads: wrong methods of feeding and care, and specific infection.

The first milk of the mother is a natural laxative and aids in ridding the intestine of the young of such waste material (meconium) as collects during fetal life. If this milk is withheld, the intestine becomes irritated, constipation occurs, followed by a diarrhoea or serious symptoms of a nervous character, caused by the poisonous effect of the toxic substances absorbed from the intestine on the nervous system.

Changes in the ration fed the mother, excitement, unusual exercise and disease change the composition of the mother's milk. Such milk is irritating to the stomach and intestines of the young. This irritation does not always develop into a diarrhoea, but may result in a congestion of the stomach.

When the young are raised artificially or by hand, and fed milk from different mothers of the same or different species, or changed from whole to skim milk, acute and chronic digestive disorders that are accompanied by a diarrhoea are common. Feeding calves from filthy pails, allowing them to drink too rapidly and giving them fermented milk are common causes of scours.

White scours caused by irritating germs is a highly infectious disease. The disease-producing germs gain entrance to the body by way of the digestive tract and the umbilical cord.

Insanitary conditions, such as dark, cold, damp, filthy quarters, lower the vitality of young animals, and predispose them to digestive disorders as well as other diseases.

The symptoms are as follows: Constipation accompanied by a feverish condition precedes the diarrhoea; colicky pains are sometimes manifested;

the diarrhoea is usually accompanied by depression, falling off in appetite and weakness. At first the intestinal discharges are not very foul smelling; later the odor is very disagreeable. The faeces may be made up largely of undigested, decomposed milk that adheres to the tail and hind parts. If the diarrhoea is severe, the animal refuses to suckle or drink from the pail, and loses flesh rapidly. It is usually found lying down. The ears droop and the depression is marked. The body temperature may vary from several degrees above to below the average normal.

The infectious form of white scours may be diagnosed by the history of the outbreak. In this form of the disease, a large percentage of the young are affected and the death-rate is very high.

Calves and lambs frequently die of an acute congestion of the fourth stomach. In this disease, the symptoms appear shortly after feeding. It is characterized by colicky pains, convulsions and coma.

The treatment is largely preventive. Young animals should be provided with dry, clean, well-ventilated quarters and allowed plenty of exercise. Colts thrive best if allowed to run in a blue grass pasture with the mother. If the mother is worked, suitable provisions in the way of quarters and frequent nursing should be provided. Calves, lambs and pigs are the most frequent sufferers from insanitary quarters. In breeding, we should always strive to get strong, vigorous, healthy young. The care given the mother in the way of exercise and feeding is an important factor here.

The first milk of the mother should not be withheld from the young, especially if the animal is raised by hand. We must also feed it regularly and not too much at any one time. Any change in the milk should be made gradually, and it is usually advisable to reduce the ration slightly when such a change is made, so as not to overwork the digestive organs. Pails and bottles from which the animal feeds should be kept clean.

Colts raised on cow's milk must be fed and cared for carefully. The milk must be sweet and made more digestible by diluting it with one-third water. A little sugar may be added. It is very advisable to add from one-half to one ounce of lime water to each pint of milk fed. Frequent feeding is very necessary at first, and we must not underestimate the quantity of milk necessary to keep the colt in good condition. It should be taught to eat grain as soon as possible.

Because of the irritated condition of the stomach and intestine, the animal suffering from diarrhoea is unable to digest its feed. For this reason it is very important to withhold all feed for at least twelve hours. Water should be provided. The alimentary tract is relieved of the irritating material by giving the animal a physic of castor or linseed oil. The dose varies from one-quarter to one-half ounce for the lamb and from one to four ounces for the colt or calf. It is advisable in most cases to follow this with the following mixture: bicarbonate of soda one ounce, bismuth subnitrate one-half ounce, and salol one-quarter ounce. The dose for the colt and calf is one teaspoonful three times a day. Lambs and pigs may be given from one-fourth to one-half the above dose.

It is usually advisable to give ewes and sows a physic if their young develop a diarrhoea. Mothers that are heavy milkers may be given a physic the second or third day following birth. The ration should be reduced as well during the first week.

DISEASES OF THE DIGESTIVE TRACT OF POULTRY

GENERAL DISCUSSION.—The digestive tract of poultry is composed of the following organs: mouth, gullet, crop, stomach, gizzard and intestines, with the two large glands, the liver and pancreas. The digestion of the feed begins in the crop. Here the feed is held for a short time, mixed with certain fluids and softened. On reaching the stomach it becomes mixed with the digestive fluid secreted by the gastric glands. This second digestive action consists in thoroughly soaking the feed in the gastric juice, making it soft and preparing it for maceration by the heavily muscled gizzard. Following maceration it passes into the intestine. It is here that the digestive action is completed and absorption occurs.

Under the conditions of domestication, poultry are subject to a great variety of intestinal disorders.

DISEASES OF THE CROP.—Impaction and inflammation are the two common diseases of the crop. *Large, impacted crops* are usually caused by the feeding of too much dry feed, fermentation of the contents of the crop and foreign bodies that obstruct the opening from the organ.

Inflammation of the crop is caused by excessive use of condiments in the feed, putrid or spoiled feeds and eating caustic drugs, such as lime and rat poison.

The symptoms are dulness, an indisposition to move about, drooping wings and efforts to eject gases and liquids. The crop is found greatly distended and either hard or soft, depending on the quantity of feed present and the cause of the distention. If fermentation is present the crop usually feels soft.

The preventive treatment consists in practising proper methods of feeding. The *curative treatment* of a recent case consists in manipulating the mass of feed, breaking it up and forcing it upwards toward the mouth. If difficulty in breaking up the mass is experienced, it is advisable to administer a tablespoonful of castor oil to the bird.

If the above manipulations are unsuccessful, an operation is necessary. This consists in making an opening through the skin and the wall of the crop and removing the contents with tweezers. The opening must be closed with sutures. The proper aseptic precautions must be observed.

In inflammation of the crop, the bird should be dieted for at least one day, and one teaspoonful of castor oil given as a laxative.

ACUTE AND CHRONIC INDIGESTION.—The recognition of special forms of indigestion in poultry is difficult. A flock of poultry that is subject to careless and indifferent care may not thrive and a number of the birds develop digestive disorders. This may be indicated by an abnormal or depraved appetite and emaciated condition. Constipation or diarrhoea may occur. In the more severe cases the bird acts dull, the feathers are ruffled and it moves about very little.

The treatment consists in removing the cause, and giving the flock a tonic mixture in the feed. The following mixture may be used: powdered gentian and powdered ginger, eight ounces of each, Glauber's salts four ounces, and sulfate of iron two ounces. One ounce of the above mixture may be given in ten pounds of feed.

WHITE DIARRHOEA OF YOUNG CHICKENS.—White diarrhoea is of the greatest economic importance to the poultryman. The loss of chicks from this disease is greater than the combined loss resulting from all other diseases. It is stated by some authors that not less than fifty per cent of the chickens hatched die from white diarrhoea.

Such a heavy death-rate as is attributed to this disease can not result from improper methods of handling and insanitary conditions. Before it was proven that white diarrhoea was caused by specific germs, a great deal of emphasis was placed on such causes as debilitated breeding stock, improper incubation, poorly ventilated, overcrowded brooders, too high or too low temperatures and filth. Such conditions are important predisposing factors, and may, in isolated cases, result in serious intestinal disorders.

The microorganisms causing this disease belong to both the plant and animal kingdoms. Infection usually occurs within a day or two following hatching. Chicks two or three weeks of age seldom develop the acute form of the disease. Incubator chicks are the most susceptible to the disorder.

The following symptoms occur: The chicks present a droopy, sleepy appearance; the eyes are closed, and the chicks huddle together and peep much of the time; the whitish intestinal discharge is noticed adhering to the fluff near the margins of the vent, and the young bird is very weak; death may occur within the first few days. After the first two weeks the disease becomes less acute. In the highly acute form the chicks die without showing the usual train of symptoms.

It is very easy to differentiate between the infectious and the non-infectious diarrhoea. In the latter, the percentage of chicks affected is small and the disease responds to treatment more readily than does the infectious form. The death-rate in the latter form is about eighty per cent.

The treatment of diarrhoea in chicks from any cause is preventive. This consists in removing the cause. No person can successfully handle poultry if he does not give the necessary attention to sanitation. Poultry houses, runs, watering fountains and feeding places must be constantly cleaned and disinfected. The degree of attention necessary depends on the surroundings, the crowded condition of the poultry houses and runs, and the presence of disease in the flock. If disease is present, we can not clean and disinfect the quarters too often. The attendant can not overlook details in handling the incubator or brooder and feeding the chicks and be uniformly successful.

If the disease is known to be present in the flock, the incubators and brooders should be thoroughly disinfected by fumigating them with formaldehyde gas. If dirty, they should first be washed with a water solution of a good disinfectant. For a period of from twenty-four to forty-eight hours after hatching, the chicks should receive no feed. Dr. Kaupp recommends as an intestinal antiseptic, sulfocarbolate thirty grains, bichloride of mercury six grains, and citric acid three grains, dissolved in one gallon of water. This solution should be kept in front of the chicks all the time. A water solution of powdered copper sulfate (about one-half teaspoonful dissolved in one gallon of water) may be used.

QUESTIONS

1. Name the organs that form the digestive apparatus.

2. What digestive action on the feed occurs in the mouth?

3. Describe the causes and symptoms of inflammation of the mouth; describe the treatment.

4. Give the causes for depraved appetite; describe the symptoms and treatment.

5. Give the capacity of the horse's stomach.

6. Name the different compartments of the ruminant's stomach.

7. Give the capacity of the stomach of ruminants.

8. Name the different stages of digestion occurring in the stomach of the hog.

9. What forms of acute indigestion involve the stomach of solipeds? Give causes and treatment.

10. Give the causes of indigestion of the stomach of ruminants.

11. Give the treatment for the different forms of indigestion of the stomach of ruminants.

12. Name the divisions of small and large intestines of solipeds and ruminants.

13. What is the capacity and length of large intestine of solipeds and ruminants?

14. What are the different forms of acute indigestion of the horses, and causes?

15. Give a general line of treatment for acute indigestion of the horse.

16. Give the causes of white diarrhoea in the young chicks; give a line of treatment.

17. Name the organs of the digestive apparatus of poultry.

18. Name the common diseases of the digestive apparatus of poultry, and give the causes.

CHAPTER V
DISEASES OF THE LIVER

GENERAL DISCUSSION.—The liver is one of the most important glands of the body, as well as the largest. Because of its physiological influence over the functions of the kidneys, intestines, and body in general and the varied functions that it possesses, it is frequently affected by functional disorders.

All of the blood that comes directly from the intestine is received by the liver. It secretes the bile, neutralizes many of the poisonous substances and end products of digestion that are taken up by the absorbing vessels of the intestine, and acts as a storehouse for the glycogen.

It can be readily understood from this brief statement of the nature of the liver functions, that any functional disorder of the liver may be far reaching in its effect. In many of the diseases that involve other organs, the liver may be primarily affected. It is difficult to diagnose functional disorders of the liver that are responsible for a diseased condition of some other body organ. A knowledge of the physiology and pathology of the liver is of the greatest importance in the diagnosis of this class of disorders.

In the larger domestic animals, symptoms of liver diseases are more obscure than in the small animals. In certain parasitic diseases and in mixed and specific infectious diseases, the liver may show marked pathological changes.

COMMON CAUSES OF LIVER DISORDERS.—Domestic animals commonly live under very unnatural conditions. Ill results do not follow unless these conditions are so extreme as to violate practically all of the health laws. Pampered animals are especially prone to liver disorders. The feeding of too heavy and too concentrated a ration together with insufficient exercise is one of the most common causes of disorders of the liver. The feeding of a ration that is unsuitable for that particular species is a common source of disease in animals. For example, the feeding to carnivora of a ration made up largely of starchy feed, and the feeding of a ration containing an excessive quantity of protein to herbivorous animals may result in intestinal, liver and nervous disorders. Spoiled feed may prove highly injurious. Catarrhal inflammation of the intestine and intestinal parasites may obstruct the bile duct, and interfere seriously with the functions of the liver.

Symptoms.—In diseases of the liver the appetite is irregular or the animal refuses to eat, is constipated, or has diarrhoea. The faeces may be grayish colored or foul smelling. Colicky pains are sometimes manifested. Usually the animal acts dull and weak. A raise in body temperature may be noted. The visible mucous membranes may appear yellowish- or brownish-red in color.

Treatment.—Animals grazing over well drained pastures that are free from injurious weeds and provided with plenty of drinking water, seldom develop diseases of the liver. Exercise, a natural diet and plenty of clean water, as well as preventing liver disorders, may be classed among the most important of all curative agents. Laxatives or cathartics, such as oils, salts, aloes, and calomel, in small doses may be given. We prefer the administration of oil or aloes to horses, Glauber's or Epsom salts to ruminants, and calomel to dogs. The administration of minimum doses of these drugs, and repeating the dose after a short interval, is preferable to large doses. Alkaline tonics are also indicated. The following mixture may be given: bicarbonate of soda, sulfate of soda and common salt, eight ounces of each, and powdered gentian and sulfate of iron, four ounces of each. Large animals may be given a small tablespoonful of this mixture with the feed three times a day. The dose for sheep and hogs is one teaspoonful. A very light, easily digested ration should be fed.

QUESTIONS

1. What can be said of the importance of the liver?

2. Tell something of its duties as a gland.

3. In what animals are liver troubles most conspicuous when present?

4. Give causes of liver disorders.

5. What are the symptoms?

6. What are the most important natural cures?

7. What rule may be given for adapting suitable laxatives to different classes of animals?

CHAPTER VI
DISEASES OF THE URINARY ORGANS

GENERAL DISCUSSION.—The urinary apparatus is composed of two glands, the kidneys and an excretory apparatus that carries the excretion of the kidneys to the outside.

The kidneys are situated in the superior region of the abdominal cavity (sublumbar) above the peritoneum, and to the right and left of the median line. They are highly vascular glands, somewhat bean-shaped and of a deep red color. These glands are capable of removing from the blood a fluid that is essentially different in composition and which, if retained in the blood, would be harmful or poisonous to the body tissues.

The kidney excretions are carried from the pelvis of the kidneys by the right and left ureters. These canals terminate in the bladder, an oval-shaped reservoir for the urine. This organ is situated in the posterior portion of the abdominal cavity and at the entrance to the pelvic cavity. Posteriorly, it forms a constricted portion or neck. It is here that the urethra originates. This canal represents the last division of the excretory apparatus. In the female, the urethra is short and terminates in the vulva. In the male it is long and is supported by the penis.

The urine secreted by the kidneys is a body excretion, and consists of water, organic matter and salts. The nitrogenous end-products, aromatic compounds, coloring matter, and mucin form the organic matter. The nitrogenous end-products and aromatic compounds are urea, uric and hippuric acids, benzoic acid and ethereal sulfates of phenol and cresol. The salts are sulfates, phosphates and chlorides of sodium, potassium, calcium and magnesium. The organic and inorganic matter varies with the ration.

The quantity of urine secreted within a given time varies in the different species and at different times in the same individual. In the horse the quantity secreted in twenty-four hours varies from twelve to fifteen pints; in cattle from ten to forty pints; in sheep from one-half to one and three-quarter pints. The normal color of the urine varies. In the horse it is yellow or yellowish-red; in cattle and sheep yellowish; and in the dog a straw yellow. The specific gravity varies with the quantity secreted and the ration fed. When the quantity of urine secreted is above the average, the specific gravity is usually low.

THE NECESSITY OF EXAMINING THE URINE.—In diseases of the urinary apparatus, a careful examination of the urine is very necessary in order to be able to form a correct diagnosis. In domestic animals it is impractical to attempt to determine the exact amount of urine passed within a certain time, but we can make a general estimate of the quantity passed by carefully observing the animal and noting the condition of the bedding in the stall. The sample of urine to be examined is best taken from urine collected at different periods during the day. We should note its color and consistency. The different substances in the urine can be determined only by determining the specific gravity, testing with certain chemical reagents and by making a microscopic examination of the sediment. Normal urine from the horse may be turbid or cloudy and more or less slimy, because of the presence of mucin. This is less true of other species. In disease the color of the urine may be changed to a pale yellow, red or brown. For example, in congestion of the kidneys the urine is light in color and rather transparent; in southern cattle fever it may be red; and in azoturia it may be brown.

EXCESSIVE URINATION.—The horse is the most common sufferer from excessive secretion of urine. The most common *causes* are musty feeds, such as hay, grain and shipped feeds. New oats, succulent feeds and acrid plants may sometimes cause it. In the fall of the year, when the season is changing from warm to cool weather and the horse eliminates less water from the body by way of the skin, the kidneys may become more active and the quantity of urine secreted be greatly increased. This, however, is a normal physiological condition and should not be confused with this disease.

The first *symptom* noted is the frequent passing of a large quantity of urine. The animal drinks more water than usual and the appetite is poor. Dulness and a weak, emaciated condition are prominent symptoms. Death occurs unless the cause of the disease is removed. If the poisonous substance has been acting for some time, it is difficult to cure the animal.

This disease can be *prevented* by eliminating spoiled feeds from the ration fed to animals in our care. Early in the attack the necessary attention to the ration and the feeding of a clean, nourishing ration is sufficient to correct the disease. The quantity of water drunk by the animal should be limited. Complete rest is indicated. Laxatives, stimulants and tonics should be given if necessary.

NEPHRITIS.—Congestion and inflammation of the kidneys commonly occur in mixed and specific infectious diseases, such as septicaemia, pyaemia and influenza. The toxic effect of spoiled feeds, impure drinking water, and irritating drugs like cantharides and turpentine may so irritate the kidneys as to cause them to become inflamed. Chilling of the skin and nervousness or extreme fear may sometimes cause a congestion of these organs. Inflammation of the kidneys is a common complication of azoturia. Irritation from parasites should be included among the causes of this disease.

The *symptoms* vary in the different stages of the disease. During the period of active congestion the quantity of urine secreted is increased. The scant secretion of urine, dark in color and thick or turbid, is suggestive of an inflammation of the kidneys. The animal moves stiffly, the back may be arched, urination is painful and the urine is passed in very small amounts. The appetite is irregular or suppressed, the pulse strong at first but later small and weak, and the body temperature is elevated. On making a rectal examination we find the bladder empty and the kidneys enlarged and sensitive.

When the kidneys become so badly diseased that they can no longer perform their function of separating from the blood the nitrogenous end-products of digestion, uraemic poisoning occurs. In this later stage of the disease the animal staggers about if moved, and finally goes down in the stall and is unable to get up. Death is usually preceded by convulsions and coma.

The prognosis is very unfavorable, death occurring in the majority of cases. In azoturia of horses and in infectious diseases, the inflammation is nearly always acute. The color of the urine, its high specific gravity and the small quantity passed are valuable symptoms to consider in the recognition of this disease. Chronic inflammation generally develops slowly and may not give rise to any very prominent symptoms at first.

The preventive treatment of nephritis consists in careful nursing of animals affected with acute infectious diseases, a clean water supply and avoiding the feeding of spoiled feeds. The *curative treatment* is largely careful nursing. The animal should be given comfortable, well-ventilated quarters and complete rest. Chilling of the skin should be especially guarded against by protecting the body with heavy blankets and applying roller bandages to the limbs when necessary. The diet must be of such a nature as not to increase the work of the kidneys. For the first few days the animal should receive very little feed or water. Later a sloppy diet of sweet milk, green feed and mashes should be fed. Such purgatives as aloes and Glauber's salts are indicated at a very early stage in the disease. We must encourage the elimination of waste products by way of the skin in the larger animals by vigorous rubbing, blanketing and the administration of such drugs as pilocarpine. If the animal becomes weak, general and heart tonics may be given.

CYSTITIS.—Inflammation of the bladder is not an uncommon disease of horses. It is commonly *caused* by retention of the urine, calculi in the bladder and chilling of the body. Irritating drugs that are eliminated from the body in the urine, and infection of the bladder by germs may cause it.

The symptoms are usually marked. The inflammation is characterized by more or less pain, depending on the degree of the inflammation, and frequent passing of urine. Only a small amount of urine is passed at each attempt, and in severe inflammation it may contain pus or blood. Colicky pains sometimes occur. The pain is usually manifested by a stiff, straddling gait and tenderness when pressure on the bladder is made by introducing the hand into the rectum or vagina, and pressing over the region of the bladder. General symptoms, such as elevation in body temperature and irregular appetite, may be manifested.

The treatment should be first directed at removing the cause. If a cystic calculus is present in the bladder it should be removed. If the retention of the urine is caused by some local condition, and this is very often the case in nervous, well-bred animals, this must first be corrected. It is best to feed green and soft feeds, such as bran mash and chopped hay, and, if the animal will take them, gruels. A physic of castor or linseed oil should be given occasionally. It is very necessary that the animal be kept quiet. Comfortable, clean quarters and a good bed should be provided. Whenever necessary the animal should be blanketed. The medicinal treatment consists in irrigating the bladder with antiseptic solutions, and administering drugs that when eliminated by way of the urine may change its composition and render it less irritating. The following mixture may be given: potassium chlorate two ounces, salol one-half ounce, and powdered nux vomica one ounce. This mixture may be divided into sixteen powders. One of the powders should be given with each feed.

RETENTION OF THE URINE.—This may be due to a variety of *causes*. In the ox and ram, small calculi collect in the S-shaped curvature of the urethra, or at its terminal extremity. In the horse, cystic calculi are more common than urethral. In cattle and hogs, fatty secretions from the inflamed lining membrane of the sheath of the male may accumulate, and obstruct the flow of urine from the anterior opening. The giving of feed rich in salts, concentrated urine resulting from feeding of too dry a ration, insufficient exercise and inflammation of the bladder are the direct causes of calculi.

Compression of the urethra by growths or tumors, strictures of the urethra, distended bladder, spasm of the neck of the bladder in nervous animals, paralysis of the bladder and injuries to the penis are common causes of retention of the urine.

The early symptoms in ruminants are not usually recognized until a day or two after retention of the urine has occurred. The symptoms are then quite marked. The animal acts dull, refuses to eat, rumination is stopped, and there is a constant effort to urinate, as indicated by the raising of the tail and rhythmical contractions of the urinary muscles just below the anus. Urine may dribble from the sheath or the flow may be completely suppressed. The odor of urine may be marked.

Horses show symptoms of abdominal pain. The animal may move about the stall, lie down and get up again, or make unsuccessful attempts to urinate. On examination the bladder is found to be greatly distended with urine. In the horse the retention is recognized at an earlier period than in ruminants, because of the prompt, decided symptom of pain.

Retention of the urine commonly terminates in rupture of the bladder in ruminants. When this occurs, the symptoms of pain are less evident. Death occurs from uraemic poisoning and peritonitis. The outcome is less favorable in ruminants than in solipeds.

Inflammation of the sheath can be readily recognized because of the local swelling.

The following lines of treatment are recommended: A ration or feed that favors the formation of calculi should not be fed to animals; inflammation of the sheath should receive prompt treatment—this consists in irrigating the part with warm, soapy or alkaline water, followed by an antiseptic wash; we may attempt to work the urethral calculi forward and out of the S-curve in the urethra; if this is unsuccessful, urethrotomy for their removal may be attempted.

The retention of the urine in horses, because of spasm or paralysis of certain muscles, may be treated by passing the catheter. Sometimes spreading litter under the horse and keeping it quiet may induce it to urinate. Hot packs over the region of the back may be used. The treatment for calculi is entirely surgical. The operation for the removal of cystic calculi in the horse, although difficult, is followed by good results.

1. Describe the urinary apparatus.

2. Give the composition of the urine and quantity secreted in the different animals.

3. State method of determining quantity and composition of urine secreted by different domestic animals.

4. Give the causes and treatment of excessive urination.

5. Give the causes and treatment of congestion and inflammation of the kidneys.

6. Give the causes of cystitis; symptoms; treatment.

7. Give the causes and treatment of retention of the urine.

CHAPTER VII
DISEASES OF THE GENERATIVE ORGANS

GENERAL DISCUSSION.—The study of the organs concerned with the reproduction of the species is essential in order to acquire a knowledge of their several functions. It is only through an understanding of these functions that we can prepare ourselves to correctly recognize, and successfully treat, or prevent, such diseases as may involve the organs of generation. A knowledge of the structure and function of the generative organs of the female is of greater importance from the standpoint of disease, than is a similar knowledge of the generative organs of the male. The female is concerned with the complete reproductive process, which may be divided into four stages. These are *copulation, fecundation, gestation* and *parturition*. The male is concerned only with *copulation* and *fertilization* of the ovum by the spermatozoa, while the female must protect and nourish the embryo and foetus until it has become sufficiently developed to live independently of the protection and nourishment afforded it within the womb. When the final stage of gestation is reached, birth or the act of parturition occurs.

GENITAL ORGANS OF THE FEMALE.—The female generative organs are the ovaries, fallopian tubules, uterus, vagina, vulva and mammary glands.

The ovaries are analogous to the testicles of the male. Their function is to secrete ova. This pair of glands is suspended from the superior region (sublumbar) of the abdominal cavity by folds of the lining membrane. Leading from the ovaries, but connected with the surface of these glands only during the period of oestrum or heat, are the fallopian tubules. Their function is to carry the ovum from the ovaries to the uterus.

The uterus or womb is a membranous sack situated in the sublumbar region and at the inlet to the pelvic cavity. It is held in position by numerous folds of the lining membrane of the abdominal cavity. We may divide the womb into three divisions, cornua, body and cervix.

The cornua or horns are long and cylindrical in shape. This portion of the womb is greatly developed in animals, like the sow and bitch, that give birth to several young. In the impregnated animal the wall of the cornua that contains one or several foetuses, and the body as well, becomes greatly thickened and the lining membrane more vascular.

The body is short in all domestic animals and connects the horns with the cervix or neck. The latter is represented by a narrow portion that projects backward into the vagina. In the cow the cervix is less prominent than in the mare and the tissue that forms it, quite firm. In the cow the opening in the cervix, the os, is very small.

The vagina is a musculo-membranous canal that leads from the womb. In the mare and cow it is about one foot in length. Its function is to take part in copulation and parturition.

The vulva is the external opening of the maternal passages. It shows a vertical slit enclosed by lips, and interiorly it forms a passage that is continuous with the vagina. This passage is about six inches long in the larger animals. The different features that should be noted are the clitoris, a small erectile organ located at the inferior portion of the opening, the meatus urinaris, the external opening of the urethra, situated in a depression in the floor of the vulva, and the hymen, an incomplete membranous partition that may be found separating the vulva from the vagina.

The mammary glands or udders secrete the milk that nourishes the young. The glands vary in number. The mare has two, the cow four , the ewe two and animals that give birth to several young, eight or more. Each gland is surmounted by a teat or nipple. The glandular tissue consists of caecal vesicles that form grape-like clusters around the milk tubules. The milk tubules from the different portions of the gland converge and form larger tubules that finally empty into small sinuses or reservoirs at the base of the teat. Leading from these sinuses are one or several milk ducts that open at the summit of the teat.

GENITAL ORGANS OF THE MALE.—The genital organs of the male are the testicles, the ducts or canals leading from the testicles, the seminal vesicles, the glands lying along the urethra, and the penis.

The testicles are the glandular organs that secrete the spermatozoa, the essential elements of the seminal fluid. These glands are lodged in the scrotal sack, situated between the two thighs.

Lying along the superior border of the testicle is a mass of ducts, the *epididymis*. The *vas deferens* is the canal or duct that passes from the epididymis to the region of the bladder and terminates near its neck by emptying into the seminal vesicles.

The seminal vesicles are two membranous pouches situated just above the bladder. They act as receptacles for the seminal fluid. Two short ducts, the *ejaculatory*, carry the seminal fluid from the seminal vesicles to the urethra.

The prostate gland is situated near the origin of the urethra. *Cowper's glands* lie along the course of the urethra and near the origin of the penis. These glands empty their secretions into the urethra and dilute the seminal fluid.

The penis is the male organ of copulation. It originates at the arch of the ischium and extends forward between the thighs. It may be divided into fixed and free portions. The free portion is lodged in the prepuce or sheath, but at the time of erection protrudes from it.

STERILITY, IMPOTENCY.—Fecundation does not always follow intercourse of the male and female. Impotency in the male and sterility in the female frequently occur.

The causes are quite varied. A normal copulation may be impossible because of injuries to, and deformities of, the parts and tumor growths. Deformed genital organs and obstructions of the os by growths and scar tissue are causes of sterility in the female.

Failure to breed is commonly caused by faulty methods of feeding and care. Over-feeding and insufficient exercise may result in the body tissues becoming loaded with fat. This may cause a temporary sterility, but if persisted in, as is frequently the case in show animals, the sterility becomes permanent because of the genital glands failing to secrete ova and spermatozoa, or the lack of vitality of the male and female elements. Old age and debility from disease or poor care may induce loss of sexual desire and an absence of, or weakened spermatozoa in, the seminal fluid. The refusal of the male to serve certain females is sometimes noted.

Tuberculosis may affect the ovaries and cause permanent sterility. In inflammation of the lining membrane of the womb and vagina, the secretions are abnormal and may collect in the womb and the passages leading to it. These secretions destroy the vitality of the spermatozoa, and this condition may be considered a common cause of sterility in the larger animals. Many vigorous young males are made impotent by excessive copulation. The excessive use of the male at any time may result in failure to impregnate a large percentage of the females that he serves.

Barren females do not become pregnant after frequent intercourse with the male. Young sterile females may not come in heat. Sometimes unnatural periods of heat are manifested, the animal coming in heat frequently or remaining in heat for a longer period than usual. This sometimes occurs in tuberculosis of the ovaries. In chronic inflammation of the maternal passage there is more or less discharge from the vulva. Both sexes may be overly fat or weakened and debilitated by disease. Deformity of the generative organs and growths may be found on making an examination. Absence of, or lack of vitality of the spermatozoa may be determined by microscopic examination of the seminal fluid.

The treatment is largely preventive. It is very important that breeding animals be kept in proper physical condition by avoiding the feeding of too heavy or too light a ration, and allowing them sufficient exercise. The male is more often affected by the latter cause than the female. This is because the average stockman does not consider exercise given under the right conditions an important factor in maintaining the vigor of the male. Young males should not be given excessive intercourse with the female. Such practice is certain to seriously affect the potency of the animal. The excessive use of the stallion can be avoided by practising artificial impregnation of a part of the mares that he is called to serve. Sterility caused by growths and closure of the os may be corrected by an operation.

Chronic inflammation of the maternal passages should be treated by irrigating the parts with a one per cent warm water solution of lysol, or liquor cresolis compound. The parts should be irrigated daily for as long a period as necessary. Fat animals should be subjected to a rigid diet and given plenty of exercise. Following this treatment a stimulating ration may be fed for the purpose of encouraging the sexual desire. In weak and debilitated animals, the cause should first be removed and a proper ration fed. Cantharides and strychnine are the drugs most highly recommended for increasing the sexual desire.

SIGNS OF PREGNANCY.—The signs which characterize pregnancy are numerous and varied. For convenience we may classify the many signs of pregnancy under two heads, probable and positive. Under the head of probable signs, we may group the following symptoms of pregnancy: cessation of heat; changes in the animal's disposition; increase in the volume of the abdomen and tendency to put on fat. The positive signs are the change in the volume of the udder; the secretion of milk; the movement of the foetus and presence of the foetus in the womb, as determined by rectal examination or by the feel of the abdomen.

The probable signs are not reliable, and should be considered only in connection with some positive sign. Persons who base their opinion of the condition of an animal that is supposed to be pregnant on probable signs, are frequently mistaken. It has frequently happened that animals whose condition was not at all certain have given birth to young, without giving rise to what may be termed characteristic probable signs.

The earliest probable symptom is the cessation of heat. In the large pregnant animals, irregular heat periods may occur, but in the majority of cases we may safely consider the animal impregnated if several heat periods are passed over.

It has been generally observed that the disposition of the pregnant animal is changed. They become more quiet and less nervous and irritable. The tendency of pregnant animals to put on fat is frequently taken advantage of by the stockman, who may allow the boar to run with the herd during the latter period of fattening.

The increase in the volume of the abdomen may be considered a *positive* sign of pregnancy in the small animals, but in the mare and cow it can not be depended on. Animals that are pregnant for the first time, do not show as great an increase in the volume of the abdomen as do animals that have gone through successive pregnant periods. The volume of the abdomen may vary greatly in the different individuals, and can not be depended on as a positive indication of pregnancy during the first two-thirds of the period of pregnancy in the larger domestic animals.

Comparatively early in pregnancy, the presence of a foetus can be determined by feeling the uterus through the wall of the rectum. In the small domestic animals the feeling of the abdomen gives the best results. In the cow this method of diagnosis is practised during the latter periods of pregnancy. The examiner stands with his back toward the animal's head, and on the right side of the cow and the left side of the mare. The palm of the hand is applied against the abdominal wall, about eight or ten inches in front of the stifle and just below the flank. Moderate pressure is used, and if a hard, voluminous mass is felt, or if the foetus moves, it is a sure sign that the animal is pregnant. It is not uncommon for the foetus to show some movement in the morning, or after the animal drinks freely of cold water. The increase in the volume of the udder occurs at a comparatively early period in animals that are pregnant for the first time. The secretion of milk and the dropping of the muscles of the quarters indicate that parturition is near. The Abderhalden test for determining whether or not an animal is pregnant is now practised.

HYGIENE OF PREGNANT ANIMALS.—Pregnant animals that are confined in a pasture that is free from injurious weeds and not too rough or hilly, and where the animals have access to clean water and the necessary shelter, seldom suffer from an abnormal birth. Here they live under the most favorable conditions for taking exercise, securing a suitable diet and avoiding injury. It may not be possible in managing breeding animals to provide such surroundings at all times, but we should observe every possible hygienic precaution, especially if the animal has reached the later periods of pregnancy.

All pregnant animals are inclined to be lazy, but, if permitted, will take the necessary exercise. Pregnant mares are usually worked. Such exercise does no harm, providing the work is not hard or of an unusual character. Cows are usually subject to more natural conditions than other domestic animals.

Protecting pregnant animals against injuries resulting from crowding, slipping and fighting is an important part of their care. Injuries from crowding together in the sleeping quarters and about feeding-troughs, or through doors and climbing over low partitions are common causes of injury in pregnant sows. Crowding together in the stable or yard, or through doorways, fighting, and slipping on floors, or icy places sometimes results in injury. It is rare, however, for cows to abort from an injury, but parturition may not be completely free from disagreeable complications. Under the conditions mentioned retention of the fetal membranes is common.

Ewes frequently suffer from too close confinement during late winter. Sows are often subject to the most unhygienic conditions. This is shown in the heavy death-rate in sows and pigs. During the late winter and early spring the conditions may be such as not to permit of exercise. Stormy, snowy, muddy weather is common at this season of the year. Persons caring for ewes and sows should see that they take sufficient exercise. It may be necessary to drive them about for a short time each day. At such times it may be advisable to give them a laxative dose of oil, or give a laxative with the feed. When there is any indication of constipation, this should be practised.

Pregnant animals should be fed carefully. We may feed animals that are not in this condition in a careless fashion, but if pregnant, over-feeding, the feeding of a fattening ration, or spoiled feed, and sudden changes in the feed can not be practised with any degree of safety. A bulky ration of dry feed and drinking impure, or too little, water may cause constipation, acute indigestion and abortion. The ration fed should contain the necessary inorganic and organic elements for the building up of the body tissues of the foetus.

At the end of the parturition period, separate quarters should be provided. The mare or cow should be given a comfortable clean stall away from the other animals. The ewe should be provided with a warm room if the weather is cold. It is always best to give the sow a separate pen that is dry and clean, and away from the other animals. All danger from injury to the mother and young should be guarded against.

ABORTION.—The expulsion of the foetus at any time during the period of gestation, when it is not sufficiently developed to live independently of the mother, is termed abortion. Abortion may be either *accidental* or *infectious*. Accidental abortion is more commonly met with in the mare and sow than the infectious form. In ruminants the opposite holds true.

The causes of accidental abortion are faulty methods of feeding and care. Injuries, acute indigestion, mouldy, spoiled feeds, chilling resulting from exposure and drinking ice-cold water, nervousness brought on by fright, or excitement and general diseases are the common causes of abortion.

Infectious abortion is most common in cows. Other domestic animals that may be affected are the mare, sow and ewe.

It is caused by a specific germ. The *Bacillus abortus* of Bang is the cause of abortion in cows, but the specific germ that produces abortion in other species of animals has not been proven. In this country, Keer, Good, Giltner and others have proven that the Bang bacillus of abortion is infectious for other species of animals, and outbreaks of this disease have been said to occur among breeding ewes pastured and fed on infected premises. Its infectiousness for the females of other species has never been proven in natural outbreaks.

The disease-producing germs are present in the body of the foetus, the fetal membranes, the discharge from the maternal passages, the faeces and milk of aborting animals. The male may carry the infection in the sheath, urethra and on the penis. The natural avenues of infection are the maternal passages and digestive tract.

It is very seldom that abortion is carried from one herd to another by means other than through the breeding of animals free from abortion to animals affected by this disease. The purchase of a bull or cow from an infected herd and breeding them to animals that are free from disease, is a common method of spreading the disease. After serving the diseased animal, the male may carry the bacillus of abortion into the maternal passages of the next cow he serves. There are numerous cases on record where the bull was a permanent carrier of the Bacillus abortus and infected nearly every animal served. The distribution of the disease in the herd following the introduction of a cow, sow, or ewe that has aborted before or after being purchased, takes place through contact of the other animals with the virus that may be present on the floor, or in the manure, or by taking the virus into the digestive tract along with the feed and drinking water. Experimental evidence indicates the latter avenue of infection.

The stallion is the most common source of infectious abortion in mares. An infected stallion may distribute the disease to a large percentage of the mares that he serves. For this reason nearly all of the mares in a certain locality may abort.

In case the infection occurs at the time of service, the abortion usually takes place during the first half of the period of pregnancy. Cows that become pregnant without recovering from the inflammation of the lining membrane of the genital tract, may abort at a very early period. McFadyean and Stockman from the artificially inoculated cases of infectious abortion in cows, showed that the period of incubation averaged 126 days.

The symptoms of accidental abortion are extremely variable. Animals that abort during the early periods of pregnancy may show so little disturbance, that the animal can be treated as if nothing had happened. During the latter half of pregnancy, and especially when the accident is caused by an injury, the symptoms are more serious. Loss of appetite, dulness, restlessness, abdominal pain and haemorrhage are the symptoms commonly noted. If the foetus is dead, it may be necessary to assist the animal in expelling it. In the latter case, death of the mother may occur.

A slight falling of the flanks, swelling of the lips of the vulva and a retention of the fetal membranes, or discharge from the vulva may be the only symptoms noted at the time abortion occurs.

The symptoms of infectious abortion vary in the different periods of pregnancy. At an early period, the foetus may be passed with so little evidence of labor that the animal pays little attention to it. The recurrence of heat may be the first intimation of the abortion. All cases of abortion are followed by more or less discharge from the vulva. This is especially true if the fetal membranes are retained. In such cases, the discharge has a very disagreeable odor. In most cases the foetus is dead. When born alive, it is weak and puny, and usually dies or is destroyed within a few days. When the attendant fails to give the animal the necessary attention, or is careless in his manipulation of the parts, inflammation of the womb, caused by the decomposition of the retained membranes, or the introduction of irritating germs on the ropes, instruments and hands, may occur. Death commonly follows this complication.

It is very important that the infectious form be diagnosed early in the outbreak. For all practical purposes we are justified in diagnosing infectious abortion, if several animals in the herd abort, especially if it follows the introduction of new animals. Methods of serum diagnosis, the agglutination and complement-fixation tests, are now used in the diagnosis of this disease.

The preventive treatment of the accidental form consists in avoiding conditions that may result in this accident. Pregnant animals should not be exposed to injuries from other animals or from the surroundings. Animals which show a predisposition to abort should not be bred. We should see that all animals receive the necessary exercise and a proper ration.

If the animal indicates by her actions that abortion may take place, we should give her comfortable, quiet quarters. It is very necessary to keep her quiet, and if restless, morphine may be given. A very light diet should be fed and constipation prevented by administering a laxative. The necessary attention should be given in case abortion occurs.

The enforcement of *preventive* or *quarantine measures* is very important in the control of infectious abortion. This is especially true of breeding herds and dairy cows. Breeders do not recognize the importance of keeping their herds clean or free from disease. It is a well-known fact among stockmen that abortion and other infectious diseases have been frequently introduced into the herd through the purchase of one or more breeding animals. Because of the prevalence of infectious abortion among cows, it is advisable to subject newly purchased breeding animals, or a cow that has been bred outside of the herd, to a short quarantine period before allowing them to mix with the herd. The breeding of cows from neighboring herds to the herd bull is not a safe practice. In communities where there are outbreaks of this disease, animals that abort, or show indications of aborting, should be quarantined for a period of from two to three months. The separation from the herd should be so complete as to eliminate any danger of carrying the disease to the healthy animals on the clothing and farm tools. If this method of control were practised at the very beginning of the outbreaks, the disease could be checked in the large majority of herds.

The foetus and membranes should be destroyed by burning. In case the animal does not pass the fetal membranes, they should be completely removed. In the cow, it is advisable to wait twenty-four hours before doing this. The animal's stall should be thoroughly cleaned and disinfected. It is very advisable to give the entire stable a thorough disinfecting. For this purpose a three or four per cent water solution of liquor cresolis compound may be used. It is advisable to apply it with a spray pump. The floor and feed troughs should be sprinkled daily with the disinfectant. All manure should be removed to a place where the animals can not come in contact with it. It is not advisable to confine the cows to a small yard. The more range they have the easier it is to control the disease.

Individual treatment is very necessary. In infectious abortion the mucous lining of the womb and the passages leading to it become inflamed. This should be treated by irrigating the parts with a warm water solution of a disinfectant that is non-irritating. This treatment should be repeated daily for a period of from two to four weeks. We must be very careful not to irritate the parts. A one-half per cent water solution of liquor cresolis compound may be used.

Animals that abort should not be bred until they have completely recovered. Small animals that have no special value as breeding animals should be marketed. Cows and mares should not be bred for a period of at least three months.

Infected males should not be used for service. The male should receive the necessary attention in the way of irrigating the sheath before and after each service.

PHYSIOLOGY OF PARTURITION.—Parturition or birth, when occurring in the mare, is designated as foaling; in the cow, calving; in the sheep, lambing; and in the sow, farrowing. A normal or natural birth occurs when no complications are present and the mother needs no assistance. When the act is complicated and prolonged, it is termed abnormal birth. The length of time required for different individuals of the same species to give birth to their young varies widely. It may require but a few minutes, or be prolonged for a day or more. The cause of this variation in the length of time required for different animals to bring forth their young, can be better understood if we study the anatomy of the parts and their functions.

Throughout the pregnant period the *expulsion of the foetus* is being prepared for. As the foetus develops there is a corresponding development of the muscular wall of the womb. The last period of pregnancy is characterized by the relaxation of the muscles and ligaments that form the pelvic walls, and a relaxation and dilation of the maternal passages. In addition, degenerative changes occur in the structures that attach the foetus to the womb, the normal structures being gradually destroyed by a fatty degeneration. This results in a separation between the fetal and maternal placenta. The contents of the womb begin to affect the organ in the same manner as a foreign body, irritating the nerve endings and producing contractions of the muscles. These contractions of the muscles help greatly in breaking down the attachments until finally the labor pains begin in earnest, and the foetus is gradually forced out of the womb, through the dilated os and into the vagina and vulva.

A normal birth is possible, only when the expelling power of the womb is able to overcome the resistance offered by the foetus and its membranes, the pelvic walls and the vagina and vulva.

The relative size of the foetus to the inlet of the pelvic cavity and its position are the most important factors for the veterinarian and stockman to consider . On leaving the womb, the foetus passes into the vagina and vulva. This portion of the maternal passages is situated in the pelvic cavity which continues the abdominal cavity posteriorly. The pelvic walls are formed by bones and ligaments that are covered by heavy muscles. As previously mentioned, the ligaments and muscles relax toward the end of pregnancy in order to prepare the way for the passage of the foetus. Before entering the pelvis it is necessary for the foetus to be forced through the inlet to this cavity. This is the most difficult part of the birth, as the bones that form the framework of the pelvis completely enclose the entrance to it. It is only in the young mother that the pelvic bones give way slightly to the pressure on them by the foetus. It can be readily understood, that when the young is large in proportion to the diameter of the pelvic inlet, it is difficult for it to pass through. This occurs when mothers belonging to a small breed, are impregnated by a sire belonging to a large breed of animals. It may also occur if the mother is fed too fattening a ration and not permitted sufficient exercise.

The part of the foetus that presents itself for entrance into the pelvic cavity and its position are of the greatest importance in giving birth to the young. Either end of the foetus, or its middle portion may be presented for entrance. The *anterior* and *posterior presentations* may be modified by the position that the foetus assumes. It may be in a position that places the back or vertebrae opposite the upper portion of the inlet, or the floor or sides of the pelvic cavity. These positions may be modified by the position of one or both limbs, or the head and neck being directed forwards instead of backwards. In the *transverse presentations*, the back, or the feet and abdomen of the foetus may present themselves for entrance to the pelvic cavity. These presentations may show three positions each. The head may be opposite the upper walls of the inlet, the foetus assuming a dog-sitting position, or it may lie on either side.

In order to overcome the friction between the foetus and the wall of the maternal passages, these parts are lubricated by the fluids that escape from the "water bags." If birth is prolonged and the passages become dry, birth is retarded. The hair offers some resistance in a posterior presentation. Young mares that become hysterical have abnormal labor pains that seem to hold the foetus in the womb instead of expelling it.

CARE OF THE MOTHER AND YOUNG.—Although birth is generally easy in the different domestic animals, it may be difficult and complicated, and it is of the greatest economic importance that special attention be given the mother at this time. It is very necessary for her to be free if confined in a stall. If running in a pasture or lot, the necessary shelter from storms, cold or extreme heat should be provided. Other farm animals, such as hogs, horses and cattle, should not be allowed to run in the same lot or pasture.

When parturition commences, the mother should be kept under close observation. If the labor is difficult and prolonged, we may then examine the parts and determine the cause of the abnormal birth. Unnecessary meddling is not advisable. Before attempting this examination, the hands should be cleaned and disinfected, and the finger nails shortened if necessary. The different conditions to be determined are the nature of the labor pains, the condition of the maternal passages, and the position and presentation of the foetus. In the smaller animals this examination may be difficult. In prolonged labor the parts may be found dry and the labor pains violent and irregular, or weak. The foetus may be jammed tightly into the pelvic inlet, it may be well forward in the womb, the head and fore or hind limbs may be directed backwards, or one or more of these parts may be directed forward in such a position as to prevent the entrance of the foetus into the pelvic inlet. Sometimes the foetus is in a transverse position. The parts that present themselves at the pelvic inlet should be carefully examined and their position determined. The necessary assistance should then be given. Any delay in assisting in the birth may result in the death of the young or mother, or both. On the other hand, unintelligent meddling may aggravate the case and render treatment difficult or impossible. There is no line of veterinary work that requires the attention of a skilled veterinarian more than assisting an irregular or abnormal birth.

The attendant must guard against infecting the parts with irritating germs, or irritating and injuring them in any way. The hands, instruments, and cords must be freed from germs by washing with a disinfectant, or sterilization with heat. The quarters must be clean in order to prevent contamination of the instruments and clothing of the attendant by filth. Extreme force is injurious. For illustration, we may take a case of difficult birth caused by an unusually large foetus. Both presentation and position are normal, the forefeet and head having entered the pelvic cavity, but the shoulders and chest are jammed tightly in the inlet, and the progress of the foetus along the maternal passages is retarded. By using sufficient force, we may succeed in delivering the young, but by pulling on one limb until the shoulder has entered the pelvis, and repeating this with the opposite limb we are able to deliver the young without exposing the mother to injury. It may be necessary to change an abnormal presentation, or position, to a normal presentation, or as nearly normal as possible. This should be done before any attempt is made to remove the foetus.

Following birth the mother should not be unnecessarily disturbed. The quarters should be clean, well bedded and ventilated, but free from draughts. If the parturition has been normal, a small quantity of easily digested feed may be fed. If weak and feverish, feed should be withheld for at least twelve hours. The mare should be rested for a few weeks. The young needs no special attention if it is strong and vigorous, but if weak, it may be necessary to support it while nursing, or milk the mother and feed it by hand. If the mother is nervous and irritable, it may be necessary to remove the young temporarily to a place where she can hear and see it, until a time when she can be induced to care for it. The principal attention required for young pigs is protection against being crushed by the mother. The cutting off and ligation of the umbilical cord at a point a few inches from the abdomen, and applying tincture of iodine or any reliable disinfectant is very advisable in the colt and calf.

RETENTION OF THE FETAL MEMBRANES.—The foetus is enveloped by several layers of membranes. The *external envelope, the chorion*, is exactly adapted to the uterus. The *innermost envelope, the amnion*, encloses the foetus. Covering the external face of the amnion and lining the inner face of the chorion is a double membrane, *the allantois*. The envelopes mentioned are not the only protection that the foetus has against injury. It is enveloped in fluids as well. Immediately surrounding it is the *liquor amnii*, and within allantois is the *allantoic fluid*.

The placenta is a highly vascular structure spread out or scattered over the surface of the chorion and the mucous membrane of the uterus, that attaches the foetus and its envelopes to the womb . It is by means of this vascular apparatus that the foetus is furnished with nourishment. The fetal and maternal placentas are made up of vascular villi and depressions that are separated only by the thin walls of capillaries, and a layer of epithelial cells. This permits a change of material between the fetal and maternal circulation. The arrangement of the placenta differs in the different species. In the mare and sow, the villi are diffused. In ruminants, the villi are grouped at certain points. These vascular masses are termed cotyledons. The maternal cotyledons or "buttons" form appendages or thickened points that become greatly enlarged in the pregnant animal.

Toward the end of the pregnant period, the attachments between the fetal and maternal placentulae undergo a fatty degeneration and finally separate. This results in contractions of the muscular wall of the uterus, and the expulsion of the foetus and its envelopes. In the mare, it is not uncommon for the colt to be born with the covering intact. This does not occur in the cow. Usually the envelopes are not expelled until a short time after birth in all animals, and it is not uncommon for them to be retained. This complication is most commonly met with in the cow.

In the mare the *retention of the fetal envelopes* or "afterbirth" is commonly due to the muscles of the womb not contracting properly following birth. Abortion, especially the infectious form, is commonly complicated by a retention of the fetal membranes. Any condition that may produce an inflammation of the lining membrane of the womb may result in retention of the "after-birth." Injuries to the uterus resulting from the animal slipping, fighting and becoming crowded are, no doubt, common causes of failure to "clean" in cows.

The symptoms are so marked that a mistaken diagnosis is seldom made. A portion of the membranes is usually seen hanging from the vulva, and the tail and hind parts may be more or less soiled. The latter symptom is especially prominent if the membranes have been retained for several days, and decomposition has begun. In such case, the discharge from the vulva is dark in color, contains small pieces of the decomposed membrane and has a very disagreeable odor. In the mare, acute inflammation of the womb may result if the removal of the "after-birth" is neglected. Loss of appetite, abnormal body temperature, weakness and diarrhoea may follow. Such cases usually terminate in death. Retention of the fetal membranes is a very common cause of leucorrhoea.

The treatment consists in removing the fetal envelopes before there is any opportunity for them to undergo decomposition. In the mare, this should be practised within a few hours after birth has occurred, and in other animals, from one to forty-eight hours. In warm stables and during the warm weather, treatment should not be postponed later than twenty-four hours. The only successful method of treatment is to introduce the hand and arm into the uterus, and break down the attachments with the fingers. In the larger animals, the use of the arm must not be interfered with by clothing. Every possible precaution should be taken to prevent infection of the genital organs with irritating germs. It is advisable in most cases to flush out the womb with a one per cent water solution of liquor cresolis compound after the removal of the fetal envelopes.

LEUCORRHOEA.—This is a chronic inflammation of the mucous membrane lining the genital tract, that is associated with more or less of a discharge from the vulva. It is common in animals that abort, or retain the "after-birth."

The discharge may be white, sticky, albuminous, and without odor, or it may be chocolate colored and foul smelling. The tail and hind parts are usually soiled with it. In chronic inflammation of the womb the discharge is intermittent. In mild cases the health of the animal is in no way impaired. Sterility is common. Loss of appetite and unthriftiness occur in severe cases.

Treatment.—Mild cases readily yield to treatment. This consists in irrigating the maternal passages with a one-half per cent warm water solution of liquor cresolis compound. This treatment should be repeated daily and continued for as long a time as necessary.

MAMMITIS.—Inflammation of the mammary gland or udder is more common in the cow than in any of the other domestic animals. In all animals it is most frequently met with during the first few weeks after birth.

A predisposing cause in the development of mammitis is a high development of the mammary glands. The following *direct causes* may be mentioned: incomplete milking, or milking at irregular intervals; injury to the udder by stepping on the teat; blows from the horns and pressure caused by lying on a rough, uneven surface; chilling of the udder by draughts and lying on frozen ground; and infection of the glandular tissue by *irritating germs*. The latter cause produces the most serious, and, sometimes, a very extensive inflammation. This form of inflammation may spread from one cow to another, causing the milk to be unfit for food, and bringing about the loss of one or more quarters of the udder.

The symptoms occurring in the different forms of mammitis differ. The inflammation may involve one or more of the glands, and may affect either the glandular or the connective tissue. In some cases the gland may appear congested for a few days before the inflammatory changes occur. The part becomes hot, swollen, tender and reddened. It may feel doughy or hard. If the connective tissue is involved (interstitial form), there is apt to be a high body temperature, the udder may be much larger than normal, is tender and pits on pressure. Loss of appetite usually accompanies this form of mammitis. Very little or no milk is secreted. Sometimes, the milk is greatly changed in appearance, is foul smelling and contains pus. In congestion of the udder and rupture of the capillary vessels, the milk may contain blood.

Mild inflammation of the udder responds readily to treatment. The interstitial form may terminate in abscesses and gangrene. The replacement of the glandular tissue by fibrous tissue in one or more quarters is not uncommon. Death seldom occurs.

The preventive treatment consists in avoiding conditions that may favor or cause an inflammation of the gland. Animals that have highly developed mammary glands should be fed a light diet just before and following parturition. Following parturition, a dose of Epsom or Glauber's salts may be given. If the young does not take all the milk, the udder should be milked out as clean as possible. Massaging the udder by kneading or stroking may be practised.

The following *treatment* is recommended: The application of a thick coating of antiphlogistin once or twice daily is a useful remedy. If the udder becomes badly swollen, it should be supported with a bandage. Extensive inflammation may be treated by the application of cold in the form of packs of cracked ice. Irrigating the gland with a four per cent water solution of boric acid is an important treatment for certain forms of mammitis. Abscess formation or suppuration should be promptly treated by opening and treating the abscesses. If gangrene occurs, it may be necessary to remove a part, or the whole of the udder.

The giving of milk discolored with blood may be treated by applying camphorated ointment twice daily.

SORE AND WARTY TEATS.—Irritation to the teats by filth, cold, moisture and injuries cause the skin to become inflamed, sore and scabby.

Preventive treatment is the most satisfactory. Sore teats may be treated by applying the following ointment after each milking: vaseline ten parts and oxide of zinc one part. Pendulous warts may be clipped off with a sharp pair of scissors. Castor oil applied to the wart daily by rubbing may be used for the removal of flat warts.

"MILK-FEVER" OR POST-PARTUM PARALYSIS.—This is a disease peculiar to cows, especially heavy milkers that are in good condition. It most commonly occurs after the third, fourth and fifth calving. The disease usually appears within the first two or three days after calving, but it has been known to occur before, and as late as several weeks after calving. The cause is not certainly known. The Schmidt theory is that certain toxins are formed in the udder, owing to the over activity of the cells of the glandular tissue.

The symptoms are characteristic of the disease. At the very beginning of the attack the cow stops eating and ruminating, becomes uneasy, switches the tail, stamps the feet, trembles, staggers when forced to walk and finally falls and is unable to get up. At first she may lie in a natural position; later, as the paralytic symptoms become more pronounced, the head is laid against the side of the body and the animal seems to be in a deep sleep . In the more severe form the cow lies on her side, consciousness is lost and the paralysis of the muscles is marked. The different body functions are interfered with; the urine is retained, bloating occurs, respirations are slow, pulse weak and temperature subnormal or normal.

Preventive treatment, such as feeding a spare diet during the latter period of pregnancy, is not always advisable. Heavy milkers should be given one-half pound of Glauber's salts a day or two before calving, and the dose repeated when the cow becomes fresh. Cows affected with milk-fever seldom die if treated promptly.

The *treatment* consists in emptying the udder by milking and injecting air or oxygen gas into the gland until it is completely distended . The milk-fever apparatus should be clean, and the air injected filtered. Before introducing the milking tube into the milk duct, the udder should first be washed with a disinfectant, and a clean towel laid on the floor for the gland to rest on. After injecting the quarter, strips of muslin or tape should be tied around the ends of the teats to prevent the escape of the air. If the cow does not show indications of recovery in from four to five hours, the treatment should be repeated.

It is very necessary to give the cow a comfortable stall and protect her from any kind of exposure. No bulky drenches should be administered. If she lies stretched out, the fore parts should be raised by packing straw under her. This is necessary in order to prevent pneumonia, caused by regurgitated feed entering the air passages and lungs. It is very advisable to give her the following mixture for a few days after the attack: tincture of nux vomica two ounces, and alcohol six ounces. One ounce of this mixture may be given four times daily in a little water.

QUESTIONS

1. Name the generative organs of the female.

2. Name the generative organs of the male.

3. Give the causes of sterility or impotency in the male and female.

4. Give the treatment of impotency in the male and female.

5. Describe the probable signs of pregnancy; positive signs of pregnancy.

6. Describe the hygienic care of the pregnant female in a general way.

7. Name the different forms of abortion; give the causes.

8. Describe the preventive treatment of infectious abortion.

9. Give a general discussion of the physiology of parturition.

10. What are the common causes of difficult birth?

11. What parts of the foetus may present themselves at the inlet of the pelvic cavity? What are the different positions of the foetus?

12. What attention should be given the mother at the time of parturition?

13. What attention should be given the young immediately after birth?

14. Give the causes of retention of the fetal membranes; state the method of removing them.

15. Give the causes and treatment of inflammation of the udder.

16. Give the cause of milk-fever; give the treatment.

CHAPTER VIII
DISEASES OF THE RESPIRATORY APPARATUS

GENERAL DISCUSSION.—The respiratory apparatus may be divided into two groups of organs, anterior and posterior. The anterior group, the *nostrils, nasal cavities, pharynx, larynx* and *trachea*, is situated in the region of the head and neck. The posterior group, the *bronchial tubes* and *lungs*, is situated in the chest or thoracic cavity.

The nostrils are the anterior openings of the air passages. The nasal cavities are situated in the anterior region of the head, and extend the entire length of the face. Each cavity is divided into three long, narrow passages by the two pairs of turbinated bones. The lining membrane is the nasal mucous membrane, the lower two-thirds or respiratory portion differing from the upper one-third, in that the latter possesses the nerve endings of the olfactory nerve and is the seat of smell. The five pairs of head sinuses communicate with the nasal cavities. Posteriorly and near the superior extremity of the nasal passages, are two large openings, the guttural, that open into the pharyngeal cavity.

The pharynx is a somewhat funnel-shaped cavity. The walls are thin and formed by muscles and mucous membrane. This is the cross-road between the digestive and respiratory passages. In the posterior portion of the cavity there are two openings. The inferior opening leads to the larynx and the superior one to the oesophagus. All feed on its way to the stomach must pass over the opening into the larynx. It is impossible, however, for the feed to enter this opening, unless accidentally when the animal coughs. The cartilage closing this opening is pressed shut by the base of the tongue when the bolus of feed is passed back and into the oesophageal opening.

The larynx may be compared to a box open at both ends. The several cartilages that form it are united by ligaments. It is lined by a mucous membrane. The posterior extremity is united to the first cartilaginous ring of the trachea. The anterior opening is closed by the epiglottis. Just within is a V-shaped opening that is limited laterally by the folds of the laryngeal mucous membrane, the vocal chords.

The trachea is a cylindrical tube originating at the posterior extremity of the larynx, and terminating within the chest cavity at a point just above the heart in the right and left bronchial tubes. It is formed by a series of cartilaginous rings joined together at their borders by ligaments and lined by a mucous membrane.

The bronchial tubes resemble the trachea in structure. They enter the lungs a short distance from their origin, where they subdivide into branches and sub-branches, gradually decreasing in calibre and losing the cartilaginous rings, ligaments and muscular layer until only the thin mucous membrane is left. They become capillary in diameter, and finally open into the infundibula of the air cells of the lungs.

The lungs take up all of the space in the thoracic cavity not occupied by the heart, blood-vessels and oesophagus. This cavity resembles a cone in shape that is cut obliquely downwards and forward at its base. The base is formed by the diaphragm which is pushed forward at its middle. It is lined by the pleura, a serous membrane, that is inflected from the wall over the different organs within the cavity. The median folds of the pleura divide the cavity into right and left portions. A second method of describing the arrangement of the pleura is to state that it forms two sacks, right and left, that enclose the lungs. The lungs are the essential organs of respiration. The tissue that forms them is light, will float in water, is elastic and somewhat rose-colored. Each lung is divided into lobes, and each lobe into a great number of lobules by the supporting connective tissue. The lobule is the smallest division of the lung and is formed by capillary bronchial tubes, air cells and blood-vessels. It is here that the external respiration or the exchange of gases between the capillaries and the air cells occurs.

VENTILATION.—It is agreed by all persons who have investigated the subject, that unventilated stable air is injurious to animals. At one time it was believed that the injurious effects resulting from the breathing of air charged with gases and moisture from the expired air and the animal's surroundings, were due to a deficiency in oxygen. It is now believed that the ill-effects are mainly due to the stagnation of air, the humid atmosphere, and the irritating gases emanating from the body excretions.

The common impurities found in *stable air* are carbonic and ammonia gas, moisture charged with injurious matter and dust from the floor and bodies of the animals. As a rule, the more crowded and filthy the stable, the more impurities there are in the air. If any of the animals are affected with an infectious disease, such as tuberculosis or glanders, the moisture and dust may act as carriers of the disease-producing germs. Infectious diseases spread rapidly in crowded, poorly ventilated stables. The two factors responsible for this rapid spread of disease are the lowered vitality of the animal, due to breathing the vitiated air, and the greater opportunity for infection, because of the comparatively large number of bacteria present in the air.

The purpose of stable ventilation is to replace the stable air with purer air. The frequency with which the air in the stable should be changed depends on the cubic feet of air space provided for each animal, and the sanitary conditions present. The principal factor in stable ventilation is the force of the wind. In cold weather it is very difficult to properly ventilate a crowded stable without too much loss of animal heat and creating draughts.

For practical purposes, the *need of ventilation* in a stable can be determined by the odor of the air, the amount of moisture present and the temperature. It is impossible to keep the air within the stable as pure as the atmosphere outside.

All dangers from injury by breathing impure air, or by draughts can be eliminated by proper stable construction, attention to the ventilation and keeping the quarters clean.

CATARRH (COLD IN THE HEAD).—Catarrh is an inflammation of the mucous membrane lining the nasal cavities that usually extends to the membrane lining of the sinuses of the head. It may be acute or chronic. The inflammation very often extends to the pharynx and larynx. Cold in the head is more common in the horse than in any of the other animals .

The most common causes of "colds" are standing or lying in a draught, becoming wet, and exposure to the cold. "Colds" are common during cold, changeable weather. Horses that are accustomed to warm stables, are very apt to take "cold" if changed to a cold stable and not protected with a blanket. Most animals are not affected by the cold weather if given dry quarters and a dry bed. Irritation to the mucous membrane by dust, gases and germs is a common cause. Influenza and colt distemper are characterized by an inflammation of the respiratory mucous membranes. In the horse, chronic catarrh is commonly caused by diseased teeth, and injuries to the wall of the maxillary sinus. In sheep, the larvae of the bot-fly may cause catarrh.

The early symptoms usually pass unnoticed by the attendant. The lining membrane of the nostrils is at first dry and red. During this stage sneezing is common. In a few days a discharge appears. This is watery at first, but may become catarrhal, heavy, mucous-like and turbid. In severe cases it resembles pus. The lining membrane of the eyelids appears red and tears may flow from the eye. Sometimes the animal acts dull and feverish, but this symptom does not last longer than one or two days unless complicated by sore throat.

Inflammation of the throat is a common complication of "colds." It is characterized by difficulty in swallowing and partial, or complete loss of appetite. Drinking or exercising causes the animal to cough. If the larynx as well as the pharynx is inflamed, distressed and noisy breathing may occur. Pressure over the region of the throat causes the animal pain.

Common "cold" terminates favorably within a week. Chronic catarrh may persist until the cause is removed and the necessary local treatment applied. Inflammation of the pharynx and larynx may persist for several weeks unless properly treated. Abscesses may form in the region of the throat. Horses frequently become thick winded as a result of severe attacks of sore throat.

The treatment is both preventive and curative. "Colds" and sore throat can be largely prevented by good care, exercise and properly ventilated stables. Mild cases require a light diet, comfortable quarters and a dry bed. Allowing the animal to inhale steam three or four times daily is useful in relieving the inflammation. Easily digested feeds, and in case the animal has difficulty in swallowing, soft feeds and gruels, should be given. The throat may be kept covered with a layer of antiphlogistin and bandaged. Glycoheroin may be given in from teaspoonful to tablespoonful doses, depending on the size of the animal. Chlorate of potassium may be given in the drinking water.

If the animal becomes run down in flesh, as sometimes occurs in chronic catarrh, bitter tonics should be given. In the latter disease, it is sometimes necessary to trephine and wash out the sinus or sinuses affected with an antiseptic solution. It may be necessary to continue this treatment for several weeks.

BRONCHITIS.—Inflammation of the bronchial tubes may be either acute or chronic. Acute bronchitis is especially common in the horse, while the chronic form is more often met with in the smaller animals, especially hogs. This disease is most common among horses during the changeable seasons of the years. It is *caused* by warm, close stables or stalls, and irritating gases emanating from the floor, or manure in the stall. In general, the causes are about the same as in cold in the head. In young animals and hogs, the inhalation of dust, and bronchial and lung worms commonly cause it. Verminous bronchitis usually becomes chronic.

In the acute form of the disease the *symptoms* come on very quickly, the fever is high and the pulse beats and respirations are rapid. Chilling of the body occurs, and the animal may appear dull and refuse to eat. The animal coughs frequently. Recovery occurs within a few days, unless complicated by sore throat and pneumonia. In the horse, bronchitis is not a serious disease, but in other animals recovery is delayed and complications are more common.

In chronic bronchitis in the horse, the animal coughs frequently, there is more or less discharge from the nostrils and the respirations may become labored when exercised. The animal is usually weak, in poor flesh and unfit for work. In other cases, symptoms of broken wind are noticed. Severe coughing spells on getting up from the bed, or on moving about are characteristic of bronchitis in hogs. Verminous bronchitis in calves and lambs is characterized by severe spells of coughing, difficult and labored breathing and a weak, emaciated condition.

The preventive treatment is the same as for "colds." In the acute form the treatment consists largely in careful nursing. Properly ventilated, clean quarters that are free from dust should be provided. The animal should be covered with a light or heavy blanket, depending on the temperature of the stable, and the limbs bandaged. A light diet should be fed for a few days. It is advisable to give the animal a physic of oil. The inhalation of steam every few hours during the first few days should be practised. Glycoheroin may be given three or four times a day.

Animals affected with chronic bronchitis should not be exercised or worked. We should guard against their taking cold, give nourishing feeds, and a tonic if necessary.

CONGESTION OF THE LUNGS.—Pulmonary congestion is generally due to overexertion and exposure to extreme heat or cold. It may occur if the animal is exercised when sick or exhausted. Hogs that are heated from exercise and allowed access to cold water, may suffer from a congestion or engorgement of the lungs. It may be present at the beginning of an attack of pneumonia or pleurisy.

The symptoms are difficult breathing and the animal fights for its breath. The body temperature may be several degrees above the normal. In the mild form, the above symptoms are not so marked. The onset and course of the disease are rapid, recovery, pneumonia, or death often occurring within twenty-four hours.

Pulmonary haemorrhage is not uncommon. The discharge from the nostrils may be slightly tinged with blood, or there may be an intermittent discharge of blood from the nostrils or mouth. The mucous membranes are pale, the animal trembles and shows marked dyspnoea.

The preventive treatment consists in using the proper judgment in caring for, and in working or exercising animals. This is especially true if the animal is affected with acute or chronic disease. At the very beginning, bleeding should be practised. Hot blankets renewed frequently and bandages to the limbs is a very necessary part of the treatment. In case of severe pulmonary haemorrhage, treatment is of little use.

PNEUMONIA.—Inflammation of the lungs is more common in horses than in any of the other domestic animals. The croupous form is the most common. The inflammation may affect one or both lungs, one or more lobes, or scattered lobules of lung tissue. The inflammation may be acute, subacute or chronic.

The causes are very much the same as in other respiratory diseases. Exposure to cold and wet, stable draughts, becoming chilled after perspiring freely and washing the animal with cold water are the common causes of pneumonia. Inflammation of the lungs is especially apt to occur if the animal is not accustomed to such exposure. Animals affected with other respiratory diseases are predisposed to pneumonia. Drenching animals by way of the nostril and irritating drenches, or regurgitated feed passing into the air passages and lungs are the traumatic causes of pneumonia.

The symptoms vary in the different forms of pneumonia. In case pneumonia occurs secondarily, the earliest symptoms are confounded with those of the primary disease. The first symptoms noticed may be a high body temperature, as indicated by chills, and refusing to eat. The visible mucous membranes are red and congested, the nostrils dilated, the respirations quickened and difficult, the expired air hot and the pulse beats accelerated. The animal coughs, and in the horse, a rusty discharge may be noticed adhering to the margins of the nostrils. The horse refuses to lie down if both lungs are inflamed. In severe cases the expression of the face indicates pain, the respirations are labored, the general symptoms aggravated, and the animal stands with the front feet spread apart. Cattle are inclined to lie down, unless the lungs are seriously affected. Hogs like to burrow under the litter.

The course of croupous pneumonia is typical, and unless it terminates fatally in the first stage, the periods of congestion, hepatization and resolution follow each other in regular manner. Auscultation of the lungs is of great value in diagnosing and watching the progress of the disease. It is more difficult to determine the character of the lung sounds in the horse and cow than it is in the small animals. This is especially difficult if the animal is fat. During the period of *congestion* which lasts about a day, one can hear both healthy and crepitating sounds. The period of *hepatization* is characterized by an absence of sound over the diseased area. The inflammatory exudates become organized at the beginning of this stage, and the air can not enter the air cells. This period lasts several days. *Resolution* marks the beginning of recovery or convalescence. Toward the end of the second period, the inflammatory exudate in the air cells has begun to degenerate. In the last stage, these exudates undergo liquefaction and are absorbed, or expelled by coughing, in from seven days to two weeks, depending on the extent of the inflammation and the general condition of the animal.

In the subacute form the symptoms are mild and may subside within a week. Sometimes *abscesses* form in the lung. *Gangrenous inflammation* of the lung can be recognized by the odor of the expired air and the severity of the symptoms. This form of pneumonia terminates fatally. If the larger portion of the lung tissue is inflamed, death from asphyxia may occur in the second stage.

The success in the *treatment* of pneumonia depends largely on the care. Properly ventilated, clean, comfortable quarters and careful nursing are highly important. Large animals should be given a roomy box stall. Cold does not aggravate pneumonia, providing the animal's body is well protected with blankets and the limbs bandaged. Wet, damp quarters and draughts are injurious. Hogs should be given plenty of bedding to burrow in. A light, easily digested diet should be fed. Very little roughage should be fed. If the animal does not eat well, it may be given eggs and milk. Weak pulse beats should be treated by giving digitalis and strychnine. Counterirritation to the chest wall is indicated. During convalescence, bitter tonics may be given. Constipation should be treated by giving the animal castor or linseed oil.

PLEURISY.—Inflammation of the pleura is most common in horses. It occurs in all farm animals and is frequently unilateral. There are two forms of pleurisy, acute and chronic. Pleuropneumonia is common when the cause is a specific germ. This occurs in tuberculosis, pleuropneumonia of horses and pneumococcus infection.

The common causes are exposure to cold, chilling winds, draughty, damp quarters, and drinking cold water when perspiring. Injuries to the costal pleura by fractured ribs and punctured wounds may cause it to become inflamed.

The early symptoms of acute pleurisy are chills, rise in body temperature, pain and abdominal breathing. The most characteristic symptom is the ridge extending along the lower extremities of the ribs (pleuritic ridge). The animal does not stand still as in pneumonia, but changes its position occasionally, its movements in many cases being accompanied by a grunt. Pressure on the wall of the chest causes the animal to flinch and show evidence of severe pain. Large animals rarely lie down. The cough is short and painful. On placing the ear against the wall of the chest and listening to the respirations, we are able to hear friction sounds. After a few days effusion occurs in the pleural cavity. Although the animal may have refused to eat up to this time, it now appears greatly relieved and may offer to eat its feed. This relief may be only temporary. If the fluid exudate forms in sufficient quantity to cause pressure on the heart and lungs and interfere with their movement, the pulse beat is weak, the respirations quick and labored, the elbows are turned out and the feet are spread apart. All of the respiratory muscles may be used. The expression of the face may indicate threatened asphyxia. We may determine the extent of the pleural exudate by auscultation. There is no evidence of respiratory sounds in that portion of the chest below the surface of the fluid. Dropsical swellings may occur on the under surface of the breast and abdomen.

In subacute cases evidence of recovery is noted in from four to ten days. *Acute pleurisy* very often terminates fatally. Under the most favorable conditions, recovery takes place very slowly, sometimes extending over a period of several months. It is not uncommon for the horse to continue having "defective wind."

The treatment consists in good care, well ventilated quarters and careful nursing, the same as recommended in the treatment of pneumonia. At the very beginning, the pain may be relieved by the administration of small doses of morphine. If the conditions in the stable permit, a hot blanket that has been dipped in hot water and wrung out as dry as possible, may be applied to the chest wall and covered with a rubber blanket. This treatment should be continued during the first few days of the inflammation. These applications may be reinforced by occasionally applying mustard paste to the sides of the chest.

The animal should be allowed to drink but a limited amount of water. The feed must be highly nutritious. Milk and eggs should be given if necessary. A laxative dose of oil should be given. Calomel, aloes, and digitalis are recommended when the effusion period approaches in order to increase the elimination of fluid, and lessen its entrance into the body cavity. If the amount of effusion is large, puncture of the thoracic cavity with a trocar and cannula may be practised. This operation should be performed carefully, and all possible precautions used against infection of the wound. During the later period of the disease iodide of potassium, iron and bitter tonics should be given.

BROKEN-WIND, HEAVES.—The terms broken-wind and heaves are used in a way to include a number of different diseases of the respiratory organs of the horse. The term heaves is applied almost wholly to an emphysematous condition of the lungs. Broken-wind may include the following diseased conditions: obstruction of the nasal passages by bony enlargements and tumors; tumors in the pharynx; enlarged neck glands; collection of pus in the guttural pouches and paralysis of the left, or both recurrent nerves (roaring).

The common causes of heaves are pre-existing diseases of the respiratory organs, severe exercise when the animal is not in condition and wrong methods of feeding. Heaves is more common in horses that are fed heavily on dusty timothy and clover hay and allowed to drink large quantities of water after feeding, than in horses that are fed green feeds, graze on pastures or receive prairie hay for roughage. Chronic indigestion seems to aggravate the disease. Over-distention of the stomach and intestines due to feeding too much roughage and grain interferes with respiration. Severe exercise when in this condition may result in over-distention, dilation and rupture of the air cells. This is the most common structural change met with in the lungs of horses affected with heaves. It is termed emphysema.

The common symptoms noted are the double contraction of the muscles of the flank with each expiration, a short, dry cough and the dilated nostrils. The frequent passage of gas is a prominent symptom in well-established cases of heaves. Chronic indigestion is commonly present in heavy horses that are not well cared for, or are given hard work. This condition aggravates the distressed breathing.

Heaves is a permanent disorder, but it may be relieved by climatic changes and careful attention to the animal's diet.

The following *preventive treatment* is recommended: Dusty hay should not be fed to horses. Clover hay is not a safe feed for horses that are worked hard. When starting on a drive after feeding, the horse should not be driven fast, but allowed to go slowly for a few miles.

The symptoms can be greatly relieved by careful attention to the diet. A limited quantity of roughage should be fed, and this should be good in quality and fed in the evening. During the warm weather, the animal should be watered frequently. After quitting work in the evening the animal may be allowed to drink as much water as it wants. Plenty of grain, soft feed and roots may be fed. A small handful of flaxseed meal given with the feed helps in keeping down constipation. Fowler's solution of arsenic may be given twice daily with the feed, in half-ounce doses for a period of ten days or two weeks. Chronic indigestion should be combated by digestive tonics.

QUESTIONS

1. Name the organs that form the anterior and posterior air passages.

2. To what conditions are the injurious effects of keeping animals in a poorly ventilated stable due?

3. State the purpose of ventilation. How can the need of ventilation be determined in a stable?

4. State the causes of "cold" in the head; give the treatment.

5. State the cause of bronchitis; give the treatment.

6. What are the causes of pneumonia? Describe the symptoms and treatment.

7. What symptoms are characteristic of pleurisy? Give the treatment for pleurisy.

8. Give the causes and treatment of "heaves."

CHAPTER IX
DISEASES OF THE CIRCULATORY ORGANS

GENERAL DISCUSSION.—The circulatory organs are the heart, arteries, veins and lymphatics. The *heart* is the central organ of the circulatory system . Its function is to force the blood through the blood-vessels. It is situated in the thoracic cavity between the lungs, and enclosed by a special fold of the pleura, the pericardial sack. There are two kinds of blood-vessels, arteries and veins. The *arteries* leave the heart and carry the blood to the many different organs of the body. The *veins* return to the heart and carry the blood from the body tissues. The *capillaries* are small blood-vessels, microscopic in size, that connect the arteries with the veins. The arteries carry the pure blood. The opposite is true, however, of the lesser or pulmonary system. The pulmonary artery carries the impure blood to the lungs, and the pulmonary veins carry the pure blood back from the lungs. The *lymphatic vessels* carry a transparent or slightly colored fluid and chyle from the tissues and alimentary canal. This system of vessels empties into the venous system.

The functions of the blood are to nourish the body tissues; furnish material for the purpose of the body secretions; supply the cells of the body with oxygen; convey from the tissues injurious substances produced by the cellular activity; and destroy organisms that may have entered the body tissues. The cellular and fluid portions of the blood are not always destructive to disease-producing organisms. In certain infectious diseases, the fluid portion of the blood may contain innumerable organisms, and destruction of the blood cells occurs.

In inflammation of tissue the circulation of the blood in the inflamed part undergoes certain characteristic changes. At the beginning there is an increase in the blood going to the part. This is followed by a slowing of the blood stream in the small vessels, and the collecting of the blood cells in the capillaries and veins. These circulatory changes are followed by the migration of the blood cells, and the escape of the fluid portion of the blood into the surrounding tissue. The character of the above circulatory changes depends on the extent of the injury to the tissue.

PALPITATION.—This disturbance in domestic animals seems to be purely functional. It may occur independent of any organic heart disease. A highly nervous condition, excitement, over-exertion, debility from disease and the feeding of an improper ration are the common causes.

The heart beats are so violent and tumultuous as to shake the body, and be noticed when standing near the animal. The heart sounds are louder than normal and the pulse beats small and irregular. It may be differentiated from spasm of the diaphragm by determining the relationship of the heart beats to the abrupt shocks observed in the costal and flank regions.

The treatment consists in keeping the animal quiet and avoiding any excitement. A quiet stall away from the other animals is best. The treatment of palpitation resulting from some organic heart disease must be directed largely at the original disease. Morphine is commonly used for the treatment of this disorder. Weak, anaemic animals should receive blood and bitter tonics. If we have reason to believe that the disturbance is caused by improper feeding, the animal should receive a spare diet for a few days. In such cases it is advisable to administer a physic.

PERICARDITIS.—Inflammation of the pericardial sack is usually a secondary disease. It is frequently met with in influenza, contagious pleuropneumonia, hog-cholera and rheumatism. Cattle may suffer from traumatic pericarditis caused by sharp, pointed, foreign bodies passing through the wall of the reticulum and penetrating the pericardial sack. The jagged ends of fractured ribs may cause extensive injury to neighboring parts, and the inflammation spreads to the pericardial sack.

The symptoms of pericarditis may not be recognized at the very beginning when the disease occurs as a complication of influenza, or infectious pleuropneumonia. The manifestation of pain by moving about in the stall, refusing to eat and the anxious expression of the face are the first symptoms that the attendant may notice. The body temperature is higher than normal, and the pulse rapid and irregular. On auscultation, friction sounds that correspond to the tumultuous beats of the heart are heard. When fluid collects within the pericardial sack, the heart beats become feeble and the pulse weak. Labored breathing and bluish discoloration of the lips follow. The disease usually runs a very acute course. The prognosis is unfavorable.

The treatment recommended in pneumonia is indicated in this disease. Absolute rest and the feeding of an easily digested, laxative diet is a very essential part of the treatment. At the very beginning morphine may be given to quiet the tumultuous beats of the heart. Cold applications to the chest wall in the form of ice packs should be used. Heart tonics and stimulants such as digitalis, strychnine and alcohol should be administered when the pulse beats weaken. To promote absorption of the exudate, iodide of sodium may be given. Mustard paste, or a cantharides blister applied over the region of the heart is useful in easing the pain and overcoming the inflammation. If fluid collects in sufficient quantity to seriously interfere with the heart action, the sack may be punctured with the trocar and cannula and the fluid withdrawn. Great care must be used to avoid injury to the heart and infection of the part.

ACUTE LYMPHANGITIS.—This is an inflammation of the lymphatic vessels of one or both hind limbs. The attack comes on suddenly and usually occurs in connection with rest, and in horses that are of slow, quiet temperament. The *exciting cause* is an infection of the part with bacteria, the infection probably occurring through some abrasion or small wound in the skin.

The local symptoms are swelling, tenderness and lameness in the affected limb. The animal may refuse to support its weight on the affected limb. The lymphatic glands in the region are swollen, and the swelling of the limb pits on pressure. In the chronic form of the disease, the regions of the cannon and foot remain permanently enlarged, and the swelling is more firm than it is in the acute form .

The general symptoms are high body temperature, rapid pulse and the partial or complete loss of appetite.

The following treatment is recommended: Exercise is indicated in cases that are not sufficiently advanced to cause severe lameness, or inability to use the limb; rest and the application of woollen bandages wrung out of a hot water solution of liquor cresolis compound are recommended; Epsom salts in one-half pound doses may be given and repeated in two or three days; a very light diet of soft feed should be given; liniments should *not* be applied until the soreness in the limb has subsided; iodide of potassium may be given twice daily with the feed.

QUESTIONS

1. What are the functions of the blood and lymph?

2. State the changes occurring in the circulation in inflamed tissue.

3. What is palpitation? Give the causes and treatment.

4. What are the common causes of pericarditis?

5. Give the causes and treatment of acute lymphangitis.

CHAPTER X
DISEASES OF THE NERVOUS SYSTEM

GENERAL DISCUSSION.—The nervous system may be divided into central and peripheral portions. The *central portion* comprises the brain or encephalon and the spinal cord. These organs are lodged in the cranial cavity and spinal canal. The nerves and ganglia comprise the *peripheral portion*. The nerves form white cords that are made up of nerve fibres. The ganglia are grayish enlargements formed by nerve cells and supporting tissue, situated at the origin of the nerve trunk or along its course.

The brain is an oval mass of nerve tissue elongated from before to behind, and slightly depressed from above to below. It terminates posteriorly in the spinal cord. It is divided into three portions: *cerebrum, isthmus* and *cerebellum*
.

The cerebrum forms the anterior portion. It is divided into two lateral lobes or hemispheres by a deep longitudinal fissure. The surface of the cerebral hemispheres is gray and roughened by pleats or folds separated by grooves or fissures. The gray or cortical layer is distinct from the white or connecting structure. The cortical layer is made up of nerve cells or areas which control the voluntary muscles of the body. It is connected with the special senses of touch, temperature and muscle-sense. The gray layer is connected with the posterior portion of the brain, the isthmus or medulla oblongata, by the white nerve tissue.

The isthmus or *medulla oblongata* is elongated from before to behind and connects the cerebral hemispheres with the spinal cord, anteriorly and posteriorly. It is divided into several different portions, and is made up largely of white connecting fibres with nuclei of gray matter scattered through them. The isthmus is hollowed out by a system of small ventricles that extend from the cerebral hemispheres to the spinal cord, where they terminate in a small, central canal. The isthmus is the highway between the spinal cord and the higher nerve centres. It has in it certain cell centres that give origin to six of the cranial nerves.

The third division of the brain is the *cerebellum*. This is a single mass supported by the isthmus. It is situated posterior to the cerebrum, from which it is separated by a transverse fold of the membranes covering the brain. This mass of nerve tissue is much smaller than the cerebrum. The white nerve tissue forms central nuclei which send out branches that ramify in every direction. The centre of the muscular sense is said to be located in this division of the brain. A second function is the maintenance of body equilibrium through its connection with the nerve of the middle ear.

The spinal cord commences at the posterior opening (occipital foramen) of the cranial cavity, and terminates posteriorly in the lumbar region at the upper third of that portion of the spinal canal belonging to the sacrum. It is thick, white in color, irregularly cylindrical in shape, slightly flattened above and below and reaches its largest diameter in the lower cervical and lumbar regions. The spinal canal is lined by the outer membrane that envelops the cord, which aids in fixing this organ to the wall of the canal. The spinal cord is formed by white and gray nerve tissue. The gray tissue is situated within the white, and it is arranged in the form of two lateral comma-shaped columns connected by a narrow commissure of gray matter. The extremities of the lateral gray columns mark the origin of the superior and inferior roots of the spinal nerves. The white tissue of the cord is also divided into lateral portions by superior and median fissures. The inferior fissure does not extend as far as the gray commissure, leaving the lateral inferior columns connected by a white commissure. There are certain centres in the spinal cord that are capable of carrying on certain reflex actions independent of the chief centre in the brain. The white matter of the cord is made up of paths over which impulses to and from the brain are transmitted.

There are twelve pairs of cranial nerves. Two pairs belong exclusively to the special senses, smell and sight. Altogether there are ten pairs that are devoted to functions connected with the head, either as nerves of the special senses or in a motor or sensory capacity (Figs. 26 and 27). There are two pairs distributed to other regions. These are the tenth and eleventh pairs. The tenth pair or pneumogastric is distributed to the vital organs lodged within the body cavities.

There are forty-two or forty-three pairs of spinal nerves given off from the spinal cord. The spinal nerves have two roots, superior and inferior. The superior is the sensory root and the inferior is the motor root, both uniting to form a mixed nerve trunk. The sensory root possesses a ganglion from which it originates.

Generally speaking, the cerebrospinal system deals with the special senses, movement of skeletal or voluntary muscles and cutaneous and muscular sensations. In addition to the above there is a distinct system termed the sympathetic. The *sympathetic system* consists of a long cord, studded with ganglia, extending from the base of the neck to the sacrum. The ganglia are connected with the inferior roots of the spinal nerves. This cord is connected with groups of ganglia and nerve fibres in the abdominal region, and this in turn is connected with terminal ganglia in distant tissues. This system of nerves is distributed to the vital organs of the body.

CONGESTION AND ANAEMIA OF THE BRAIN.—In congestion of the brain, the blood-vessels distributed to the nerve tissue become engorged with blood. It may be either active or passive.

The cause of anaemia of the brain is an insufficient blood supply. This may be due to an abundant haemorrhage and cardiac weakness caused by shock or organic heart disease.

The causes of congestion of the brain are faulty methods of care and feeding. It sometimes occurs when horses are shipped in poorly ventilated cars, or kept in close stables. Climatic changes, or changing the stable and feed, may cause it. Extremely fat animals and animals that are rapidly putting on fat are predisposed to this disorder. Improper methods of feeding, lack of exercise, constipation and excitement are the most common causes. Passive congestion may result from pressure on the jugular vein by obstructing the flow of blood from the brain, and raising blood pressure in the blood-vessels of the brain. It is sometimes caused by organic heart trouble.

The symptoms come on very suddenly in congestion of the brain. The disease may manifest itself as soon as the animal is moved out of the stall or bed, or it may come on while it is feeding. In slight cases, the animal appears excited and restless, the eyes are bright, the pupils are dilated, and the pulse beats and respirations quickened. If the animal is moving about, it may stop suddenly and show marked symptoms of a nervous disorder, such as turning around, running straight ahead and falling down. The period of excitement is usually brief and may be followed by marked depression. The mucous membranes of the head are a deep, red color.

The symptoms in anaemic conditions of the brain are loss of consciousness, stumbling, falling to the ground and sometimes convulsions. The pig and dog may vomit. Favorable cases return to the normal within a few hours. Acute inflammatory diseases of the brain and its coverings are associated with cerebral hyperaemia or congestion.

The treatment of mild cases is to give the animal quiet, well-ventilated quarters, where it can not injure itself. The animal should be first subjected to a severe diet and later given easily-digested feed. If it appears greatly excited, bleeding should be practised. Cold applications to the head should be used in all cases in the small animals. For internal treatment, purgatives are indicated. In cases of anaemia, stimulants, vigorous massage, artificial respiration and injection of physiological salt solution are indicated.

SUNSTROKE AND HEATSTROKE.—Most writers make no distinction between heatstroke and sunstroke. The latter is caused by the direct rays of the sun falling on the animal, and the former from a high temperature and poor circulation of air in the surroundings. Under such conditions, the physical condition of the animal and exertion play an important part in the production of the nervous disturbance.

The first symptoms usually noted are rapid, labored breathing, depression and an anxious expression on the face. The horse usually stops sweating. The body temperature is extremely high, the pulse beats weak, the animal trembles, falls to the ground and dies in a convulsion. Unless measures directed toward relief of the animal are taken early in the attack, death commonly occurs. Overheating is rather common in horses that are worked hard during the extremely warm weather. Horses that have been once overheated are afterwards unable to stand severe work during the hot months of the year. Horses in this condition become unthrifty, do not sweat freely and pant if the work is hard and the weather is warm.

The preventive measures consist in not exposing animals that are fat, or out of condition to severe exercise if the day is close and hot, especially if they are not accustomed to it. When handling or working animals during hot weather all possible precautions to prevent overheating should be practised.

The treatment consists in placing the animal in a cool, shady place and fomenting the body with cold water. The cold packs or cold fomentations should be applied to the head and forepart of the body only. Small doses of stimulants may be given.

MENINGO-CEREBRITIS.—The discussion of inflammation of the brain and its coverings can be combined conveniently, as the causes, symptoms and treatment vary but little. This disorder is met with in all species of domestic animals, but it is most common in horses and mules. Some writers state that meningo-cerebritis is more common during the warm season than it is in the winter. However, this does not hold true in all sections. In the middle west, this disease is more common in late fall and winter.

It is commonly caused by taking into the body with the feed and water certain organisms and toxins that are capable of producing an inflammation of the brain. The infectious organism or toxins are taken up by the absorbing vessels of the intestines.

The secondary form of the disease usually occurs in connection with other diseases such as influenza, tuberculosis and acute pharyngitis, or as a result of wound infection. Unhygienic conditions, as unsanitary and poorly ventilated stables and filthy drinking places, play a very important part in the production of the simple or acute form of meningitis.

Sudden changes in the feed and the feeding of rotten, mouldy feeds may cause it. In the fall and winter it may follow the feeding of too heavy a ration of shredded fodder or any other dry feed. Other exciting causes are overexertion, changes in climate, excitement, injuries to the head and the feeding of too heavy and concentrated a ration.

The symptoms vary in the different individuals, but in general they are the same. At first the animal is dull, or extremely nervous and sensitive to sounds. The pupils of the eye are unevenly contracted at first, later dilated. The eyes may appear staring, or they are rolled about, so that the white portion is prominent. The unusual excitement is manifested in different ways by the different species. During the dull period the animal is indifferent to its surroundings. When it is excited, the pulse beats and respirations are accelerated. The body temperature is often elevated early in the disease. There is a partial or complete loss of appetite. Paralysis may be the most prominent symptom. The animal lies in a natural position, or stretched out and lifting the head occasionally and moving the limbs, but it is unable to rise. Loss of sensibility may gradually progress until the animal becomes semiconscious, or comatose.

In case the inflammation is acute and involves the greater portion of the brain and its coverings, death occurs within a few days. Occasionally the animal survives several weeks. There are few permanent or complete recoveries.

The principal lines of treatment are preventive measures and careful nursing. This is one of the diseases that can be largely prevented by observing all possible sanitary precautions in caring for animals. It is admitted by writers that the greater majority of cases of inflammation of the brain and its coverings are caused by infection. Proper stable construction, ventilation and disposal of the manure, an occasional disinfection of the stable, cleaning and disinfecting the drinking places and water tanks, and the necessary attention to the ration greatly reduce the loss from this disease.

The animal should be gotten into a dark, quiet, roomy stall that is well bedded. A swing may be placed under a large animal if it is able to support any of its weight, and there is no evidence of nervous excitement. We should do nothing to disturb it. If possible, the position of the animal that is unable to get up should be changed, and the bed kept clean and dry.

Cold in the form of wet or ice packs should be applied to the head during the acute stage. Symptoms of excitement must be overcome by large doses of sedatives. Iodide of potassium and strychnine may help in overcoming the paralysis. The bowels should be emptied by giving an occasional physic. A very light, easily digested diet should be fed.

PARTIAL OR COMPLETE PARALYSIS OF THE POSTERIOR PORTION OF THE BODY.—This disorder is especially common in the small animals. The hog is most frequently affected.

The following causes may be mentioned: Inflammation of the spinal cord commonly occurs in influenza, strangles and mixed infections; constipation brought on by improper feeding and insufficient exercise is a predisposing cause; injuries such as strains and blows in the region of the back may also cause it; compression of the spinal cord by the vertebrae is no doubt a very common cause; dislocation, enlargement of the disks between the vertebrae, bony enlargements resulting from strains and injuries, rickets, tuberculosis and actinomycosis and tumors commonly cause compression of the cord. It is rarely caused by parasites. Young, fat animals are especially prone to injuries in the region of the back. Such animals may suffer from malnutrition of the bones, and complete fractures of the thigh bones may occur. Extreme heat from the sun's rays and close, hot quarters are probable causes.

The symptom that is most prominent is the partial or complete loss of control over the movements of the hind parts. The appetite may be little interfered with. The animal may sit on the haunches, with the limbs projecting forward, or swing the hind quarters from side to side in walking or trotting. Irregularity in the animal's movements is especially noticeable when turning or backing. In case the animal suffers pain, the spine is held rigid or arched, and when forced to move, marked evidence of pain occurs. There may be a decrease or increase in the sensibility of the part. The increase in sensibility is noticed on striking the muscles with the hand or rubbing the hair the wrong way. Spasmodic twitching or contractions in the muscles sometimes occur. There is frequent elevation of temperature. The animal is unable to pass urine or faeces, or there may be an involuntary passage of the body excretions.

The outcome of this disease is unfavorable. Acute inflammation of the covering of the cord may subside within a few days. Cases that do not recover within a few weeks should be destroyed. Paralysis of the hind parts should not be confused with rheumatism, azoturia and other disorders that may interfere with the movements of the posterior portion of the body.

The treatment is largely along preventive lines. A predisposition toward rickets, and injuries, may be prevented by feeding a proper ration, and permitting the animal to take exercise. The quarters and the attendant are frequently responsible for injuries. If this is the case, the rough handling of the animals should be immediately corrected, and any condition of the quarters that favors the crowding or piling up of animals should be changed. Large animals may be placed in swings if they are able to support a part of their weight on the hind limbs. This is especially indicated at the very beginning of the disorder. Small animals should be given a good bed. A very light, easily digested ration should be fed. An occasional physic should be administered. Strychnine and iodide of potassium may be given. Cold applications to the back are indicated.

QUESTIONS

1. What organs comprise the central portion of the nervous system? Peripheral portion?

2. Give a general description of the brain.

3. Give a general description of the spinal cord.

4. What is the sympathetic system?

5. Describe the causes and symptoms of congestion of the brain.

6. What is heatstroke? Give the treatment.

7. Give the preventive and curative treatment of inflammation of the brain.

8. State the causes, and give the proper treatment of paralysis of the posterior portion of the body.

CHAPTER XI
DISEASES OF THE SKIN

GENERAL DISCUSSION.—The two layers that form the skin are the *epidermis* and the *derma*. The cells of the outer layer or epidermis are of two kinds. The superficial portion is formed by horny, flattened cells and the deeper by softer cells. This layer of the skin varies greatly in thickness in the different species. The derma is composed of some muscular fibres interwoven with the connective-tissue fibres. It contains the roots of the hair follicles, sweat and oil glands. The external face which is covered by the epidermis shows a multitude of little elevations. These are the vascular and nervous papillae. In addition, it shows openings through which the hairs and the skin glands pass. The inner surface is united more or less closely to the muscular or underlying tissue by a layer of fibro-fatty tissue.

The appendages of the skin are the hairs and horny productions. The horny productions comprise the horns, chestnuts, ergots, claws and hoofs.

The hair varies in length, thickness and coarseness in the different species, and the different regions of the body. In addition, breeding, care, heat and cold may cause marked variations in the thickness of the coat. Exposure to cold causes the coat to thicken. High temperatures cause the short hairs to drop out and the coat to become thin.

Diseases of the skin may be *classified as parasitic* and *non-parasitic*. Parasitic skin diseases are caused by animal and vegetable parasites. Non-parasitic skin diseases are caused by irritation to the skin and internal causes. Irritation to the skin may be either chemical, thermic or mechanical. The internal causes may be due to an individual predisposition together with digestive disturbances and the eating of feeds too rich in protein. In this chapter parasitic skin diseases produced by insects will not be discussed.

FALLING OUT OF THE HAIR AND FEATHERS.—Falling out of the hair and feathers frequently occurs independent of parasitic diseases. This condition does not occur as an independent disorder, but as a secondary affection. It is due to faulty nutrition, and irritation to the skin. Intestinal diseases, insufficient feed and feed of bad quality are common causes. Animals that are fed a heavy ration, or that lie on dirty, wet bedding frequently lose large patches of hair. Sheep that are dipped in late fall and early winter, or exposed to wet, cold weather may lose a part of their fleece. It is not uncommon for animals toward the latter period of pregnancy, or that sweat freely, to lose patches of hair.

Falling out of the hair heals of itself within a few weeks.

The preventive measures are of special importance in sheep and horses. This consists in avoiding conditions that may lead to alopecia and in correcting the diet. In horses the regions of the mane and tail should be washed with soap, or rubbed with alcohol and spirits of camphor, equal parts. Treatment should be persisted in for a long period if necessary.

URTICARIA, "NETTLERASH."—Urticaria is characterized by roundish elevations that appear quickly and become scattered over a part or the whole surface of the skin. They are caused by an inflammatory infiltration of the deeper layers of the skin. Horses and hogs are most frequently affected.

The causes of urticaria are irritating juices of certain plants, secretions of flies, ants and some caterpillars, irritating drugs, scratching, sweating and the action of cold on a warm skin. It has been observed in connection with the feeding of certain leguminous feeds and digestive disturbances. Horses that are fat, or putting on flesh rapidly, seem to be predisposed to this disorder. Urticaria may occur in certain infectious diseases.

The characteristic symptom is the formation on the skin of roundish elevations or "hives." There may be an elevation of body temperature and partial loss of appetite. Small animals may act restless and show evidence of itching or pain. This symptom is very common in hogs. The eruption may last only a few hours or a few days, or, because of the animal's scratching or rubbing the part, the skin may become scabby and small pustules form.

An important *preventive measure* is to avoid the use of agents capable of irritating the skin and producing urticaria when treating parasitic skin diseases. It is very advisable to give the animal a saline cathartic (Epsom or Glauber's salts). The skin may be washed with cold water, or a weak water solution of permanganate of potassium.

ACNE, "SUMMER RASH."—In this skin disease the oil glands and hair follicles are inflamed and sometimes infected with pus germs. This results in skin eruptions varying in size from the point of a pin to about a quarter of an inch in diameter. This inflammation is most prominent during the warm weather.

The causes are local irritation to the skin from lying on filthy floors, sweating and irritation from the harness. According to some writers, pus germs are the only cause, the mechanical agents merely aiding in the production of the infection.

The face, side of the neck, shoulders, back and sides of the trunk and quarters are the usual seats of disease. The pimples or nodules may disappear within a few weeks, or persist throughout the warm season. The eruption may disappear without leaving scars, or suppuration occurs and small bald spots result.

The treatment consists in removing the cause of the disease and cleaning the skin with antiseptic washes. The surroundings of the animal must be kept clean and a good bed provided. If possible, the horse should be laid off from work as soon as the condition is noted. Washing the part with a weak water solution of permanganate of potassium may be practised daily. Fowler's solution of arsenic may be given. This may be given with the feed.

ECZEMA.—This is an inflammation of the vascular capillary bodies and the superficial layer of the skin. There may be marked inflammatory exudate, causing the surface of the skin to become excessively moist and more or less itching. Redness, vesicles and pustules may characterize the inflammation. In the chronic form the skin may become thickened and greatly changed in structure.

Eczematous inflammation of the skin may occur in all domestic animals, but it is most common in the dog. In the horse local eczema (scratches) is common.

The most frequent *cause* is external irritation. Accumulations of filth on the skin and continual wetting of the part are common causes. Mechanical causes are rubbing, pressure, the action of the sun's rays and chemical irritants. Internal causes, such as catarrhal diseases of the stomach and weakness and emaciation from disease, may act as direct or predisposing causes. Tender-skinned animals seem to be predisposed to the disease.

The symptoms vary in the different species of animals. In the horse the thin skin posterior to the fetlock and knee, in front of the hock and on the under side of the body is most commonly inflamed. Moisture and dirt seem to be the most common causes. Eczema may involve the skin covered by the mane and tail in animals that are not properly groomed and inclined to rub or scratch. Cattle may suffer from eczematous inflammations in the region of the forehead, back of neck and base of tail. A very common form of the disease involves the space between the toes. Sheep frequently suffer from inflammation of the skin over the fetlock region. The skin of animals having long fleeces, or heavy coats of hair that become wet at a time when there is no opportunity to dry out quickly, may become inflamed. Dogs are commonly affected by the acute and chronic forms of eczema. Eczema of swine is limited mostly to young hogs. It is rather rare, excepting in hogs that are pasturing on certain kinds of clover and rape, or on muck lands.

The inflammation is accompanied by a marked tenderness and itching, and the animal licks and scratches the part. This increases the extent of the skin lesions. The skin appears moist, later dirty, scabby and thickened. Cracks and pustules may form. Gangrene and sloughing of the skin may occur.

The treatment is both preventive and curative. Cases of eczema caused by filth and wetness can be prevented by giving the necessary attention to keeping the skin clean and not allowing animals access to muddy, filthy places. Keeping the bed clean and regulating the diet are important preventive measures. Before the inflammation can be successfully treated the cause must be removed.

In acute eczema it is advisable to protect the part against water, filth and air. Powders and ointments may be used during the early stages of the inflammation. Two parts boric acid, four parts flour, and one part tannic acid may be dusted over the moist surface. One part zinc oxide and twelve parts vaseline is a useful ointment. Scratching the part should be controlled in every case by muzzles, collars and bandages. Dirt and scales may be removed from the skin by washing with cotton soaked in lime water or linseed oil. The animal should receive laxative doses of Glauber's salts or oil every few days. A simple, easily digested ration should be fed. The following mixture may be applied in obstinate cases: oil of tar and soft soap, two parts each, and alcohol one part.

COMMON FEED RASHES.—This title includes inflammation of the skin caused by pasturing on buckwheat, certain clovers and rape, together with moisture and sunlight.

Green, flowering buckwheat is more dangerous as a feed for stock than is the grain or straw. Clovers and rape are not as dangerous a feed. The actual cause of the skin becoming inflamed is not known.

The skin in the regions of the face, ears, neck, lower surface of the body and limbs becomes red and covered with vesicles. Later, scabs and pus may form.

The treatment consists in changing the ration and keeping the animals out of the sun, or long grass and weeds for a few days. This is all the treatment required in most cases. It may be advisable to administer a physic. If pus and scabs form, the part should be cleansed daily with a one per cent water solution of permanganate of potassium.

HERPES (FUNGOUS SKIN DISEASE).—This is a contagious disease of the skin caused by thread fungi, *Tricophyton tonsurans* and *epilans*, which develop in the skin in localized areas, causing vesicles, scabs or scales to appear, and the loss of the hair over the part. This skin disease occurs in all domestic animals, but it is most commonly met with in cattle. It usually affects young cattle. It most commonly occurs in the region of the face and neck. Thick, bran-like crusts form over the scattered areas of the skin and the hair drops out or breaks off. The animals frequently rub the infected area.

Prompt treatment may prevent the spread of this disease in the herd. It may be checked by quarantining the infected animals and scrubbing the stalls, stanchions and walls with a disinfecting solution. Grooming the infected animal should be discontinued. This skin disease responds most readily to ointments. Flowers of sulfur one part and lard ten parts is commonly used by stockmen. Sulfur-iodide ointment, or tincture of iodine may be applied.

QUESTIONS

1. Give a general description of the skin.

2. Give the causes and treatment of falling of the hair.

3. What is urticaria? Give the treatment.

4. What is summer rash? Give the treatment.

5. What is "scratches"? Give the treatment.

6. What feeds produce rashes of the skin?

7. What fungus produces an inflammation of the skin in cattle? Give the treatment.

CHAPTER XII
DISEASES OF THE EYE

GENERAL DISCUSSION.—*The eye* is situated in the orbital cavity, to which it is attached by muscles that rotate it in different directions. The *orbit* is lined by fibro-fatty tissues that form a cushion for the eye. Anteriorly it is protected by the *eyelids,* and in birds by a third eyelid that corresponds to the membrana nictitans of quadrupeds. The *lachrymal gland* which secretes the tears keeps the above parts moist.

The eye is the essential organ of vision. It is formed by a spherical shell which encloses fluid or semisolid parts. The shell is anteriorly made up of a transparent convex membrane, the cornea, while the remainder of its wall is formed by three opaque layers or tunics.

The external tunic is the *sclerotic.* It is a white, solid membrane, forming about four-fifths of the external shell. Its external face is related to the muscles and fatty cushion. It receives posteriorly, a little lower than its middle portion, the insertion of the *optic nerve,* which passes through the shell and spreads out to form a very thin membrane, the retina or internal coat.

The retina lines about two-thirds of the posterior portion of the shell of the eye. It is made up of seven layers. The essential layer is named from its appearance, rods and cones.

The middle coat is the *choroid.* This is a dark, pigmented, vascular and muscular membrane. The posterior portion is in contact with the retina. Anteriorly it forms the ciliary processes and the iris.

The media of the eye are the crystalline lens, vitreous and aqueous humors. The *crystalline lens* is a transparent, biconvex body sustained by the ciliary processes. The *vitreous humor* is a transparent jelly-like substance that fills all the cavity of the eye posterior to the lens. The *aqueous humor* is a liquid, contained in the anterior and posterior chambers of the eye in front of the lens. This fluid separates the iris from the front of the lens.

EXAMINATION OF THE EYE.—In examining and treating the eye we should avoid rough and hasty manipulation. The animal should be approached slowly. It is best for the attendant who is familiar with the animal to hold it for the examiner. It is advisable on approaching the animal to stroke its face, and in the horse to brush its foretop away. The hand should be carried slowly to the front of the eye, and the lids separated with the fingers and thumb if we wish to obtain a better view of the cornea. In cattle the best view of these parts can be obtained by taking hold of the nose and lifting the head. It is impossible to make a satisfactory examination of the eye outside of the stable where the light is coming from all directions. The most satisfactory conditions under which a general examination can be made is to stand the animal facing a transom, window or open door. We may then look directly into the eye and note the condition of the different refracting media.

The lens should appear transparent and free from scars. The aqueous humor free from any cloudiness or precipitate. Both pupillary openings should be the same size, and not too small or too large in the bright light. As we look through the pupillary openings, both the lens and the vitreous humor should refract the light properly and not appear white or greenish-white in color. The color of the iris should be noted. If it lacks lustre or appears dull, this may indicate an inflammation. In periodic ophthalmia in horses the iris loses its lustre and becomes a rusty-brown color. It is very important to note this change in the appearance of the iris. We should note, in addition, the expression of the animal's face, the position of the ears and eyelids and manner of the walk. Horses that have defective sight may show a deep wrinkle in the upper eyelid when startled or looking directly at an object. Animals that are blind hold the ears in a characteristic position, and may stumble and walk over, or run into objects unless stopped. The ophthalmoscope is a very useful instrument for determining the condition of the different structures of the eyes, when in the hands of persons who are trained in its use.

CONJUNCTIVITIS.—This is an inflammation of the mucous membrane lining the eyelids and covering the eyeball. The two forms of conjunctivitis common in domestic animals are the *catarrhal* and *purulent*.

The symptoms differ in the two forms of conjunctivitis. They may be distinguished from each other by the difference in the character of the inflammatory discharge. In the catarrhal form, there is a discharge of tears and the lids are held more or less closed. The mucous membrane is usually brick red in color and swollen. A little later the discharge becomes heavier and adheres more to the margins of the lids. The lids continue tender and the inflammation painful. The surface of the cornea may appear white and the blood-vessels prominent, but it is only in the severe cases that inflammation of this portion of the eye occurs. In such cases an elevation in body temperature may occur. This is especially true of purulent conjunctivitis when primarily caused by an infectious agent. In the purulent form the discharge is heavy and pus-like.

The treatment is both preventive and curative. The first object must be to remove the cause. Irritating gases resulting from stable filth should be remedied by correcting the unsanitary conditions in the stable. Conditions favoring injury to the eye from foreign bodies, such as chaff and a careless attendant, should be corrected. Animals suffering from the infectious or purulent form of inflammation should be separated from the other animals. Foreign bodies should be removed promptly before they have had an opportunity to set up a serious inflammation. It is necessary to confine the animal in some way before attempting to do this. Horses should be twitched, cattle held by the nose, and the head of a small animal held firmly with the hands. It may be necessary to cocainize the eye before the operator can remove the foreign object with absorbent cotton or with forceps.

In case of injuries and irritation to the lids by foreign bodies, the eye may be flooded with a three per cent water solution of boric acid twice daily, or as often as necessary. Such washes or lotions may be applied with a small piece of absorbent cotton, using a fresh piece each time the eye is dressed. A medicine dropper may also be used. A lotion containing silver nitrate two to four grains and distilled water one ounce, is useful in combating the inflammation. This may be applied twice daily. Irritating lotions should be avoided, if possible, in the treatment of eye diseases of horses, because of the danger of making the animal disagreeable to handle. Boric acid may be dusted over the ball of the eye of cattle with a powder blower.

PERIODIC OPHTHALMIA, "MOONBLINDNESS."—This is a periodic inflammation of one or both eyes of the horse. The internal structures of the eye are involved by the inflammation, but it may appear as a conjunctivitis.

The cause of this disease is not well understood. Certain local conditions seem to favor its development. Undrained land, a humid climate, the feeding of a one-sided ration or one that does not maintain the vitality of the animal, and severe work seem to produce it. Heredity must be accepted as a prominent accessory cause. A number of different bacteria have been mentioned as causative factors for this disease.

The symptoms at the very beginning indicate a general inflammation of the eye. The eyelids are swollen, there is an abundant secretion of tears, the eyeball is retracted and the lids are held more or less closed. As the inflammation progresses, the cornea becomes milky in appearance and the aqueous humor may show a precipitate toward the bottom of the anterior chamber. The pupil is usually contracted and dilates slowly when the animal is moved into the light. The acute inflammation gradually subsides, and about the tenth to the fourteenth day the lids and cornea may appear normal.

The periods between these acute attacks of ophthalmia may vary from a few weeks to several months. Severe work, debility and the character of the ration influence their frequency. It is not uncommon for animals that have been given a rest to suffer from a second attack on being put to work. The attendant may observe a hazy or whitish condition of the margin of the cornea. The upper lid may show an abrupt bend of its margin and a deep wrinkle. The color of the iris appears to have lost its lustre, and the aqueous humor and lens may be cloudy. After a variable number of attacks glaucoma or cataract develops.

The history of the case will enable the attendant to recognize this form of ophthalmia.

Treatment is unsatisfactory. Preventive measures consist in avoiding conditions favorable to the production of the disease. This should be practised so far as possible. At the time the attack occurs, the animal should be given a cathartic. One pound of Glauber's salts in a drench is to be preferred. Rest in a darkened stall is indicated. An eye lotion containing three grains of silver nitrate in one ounce of distilled water should be applied to the eye three times daily. A water solution of atropine or eserine should be used for the purpose of relieving the symptoms of iritis or glaucoma. A very light diet should be fed.

INFECTIOUS OPHTHALMIA OF RUMINANTS.—This occurs as an acute inflammation of the eyelids and cornea. The disease is highly infectious, affecting all of the susceptible animals in the herd. It commonly occurs during the late summer and fall.

The symptoms appear suddenly. The animal is feverish, the eyes closed and the cheeks are wet with tears. The cornea becomes clouded, white and opaque. In severe cases, the blood-vessels around the margin of the cornea become prominent, and ulcers form on its surface. The animal's appetite is impaired or lost. There is loss of flesh and temporary blindness. The blindness in one or both eyes may persist for a period of from two weeks to several months. Permanent blindness is comparatively rare.

The preventive treatment consists in practising the necessary precautions against the introduction of the disease into the herd, and in carefully quarantining the first cases of the disease that appear. The affected animal should be given a darkened stall, and fed a very light ration until the acute inflammation has subsided. From one to one and one-half pounds of Glauber's salts should be given. The *local treatment* consists in the application of antiseptic lotions or powders to the eye. Equal parts of boric acid and calomel, dusted into the eye twice daily with a powder blower, is a very effective treatment.

QUESTIONS

1. Name the different structures that form the shell of the eye; name and describe the different media of the eye.

2. Give the general method of examining the eyes of horses.

3. What is conjunctivitis? Give causes and treatment.

4. What is "moonblindness"? Give the symptoms.

5. Describe the symptoms of infectious ophthalmia of ruminants and the treatment.

CHAPTER XIII
GENERAL DISEASES OF THE LOCOMOTORY APPARATUS

GENERAL DISCUSSION.—The movements of the different parts of the animal body depend on the union of the bones that form the skeleton , and mode of insertion of the muscles. The bones meet and form *joints* or *articulations*. These are divided into three classes: *movable, mixed* and *immovable*. Nearly all of the articulations in the extremities belong to the movable class. The articulations between the bodies of the vertebrae belong to the mixed, and those between the flat bones of the head to the immovable class.

The bony surfaces that meet and form the different types of articulations are held together by ligaments . Sometimes the ligament is placed between the bony surfaces, but usually it is attached to the margins of the articular surfaces that it unites. The *immovable class* possesses fibrous-like ligaments that are placed between the margins of the flat bones that form the articulation. The *mixed articulations* are united by a fibro-cartilaginous pad that is firmly attached to the articular faces of the bones, and by peripheral ligaments that may be flat or formed by scattered fibres. All *movable articulations* are formed by bony surfaces encrusted with a thin cartilaginous layer that makes them perfectly smooth, ligaments and complimentary cartilages. Sometimes the bony surfaces do not fit each other, and we find between them *fibro-cartilages* that complete the articulation by adapting the articular surfaces to each other. *Round* or *flat ligaments* may extend from one articular surface to the other, and attached to the margins of the articulation are *membranous, flat* or *round ligaments*. Muscles and tendons that cross the articulations should be included among the structures binding them together.

Movable joints possess a *synovial membrane*. This membrane lines the structures that enclose the articulation and secretes a fluid, *the synovia*, that lubricates the surfaces.

The muscles are the contractile organs that move the body. The movement of the different parts of the body is rendered possible through the manner in which the skeletal muscles are inserted into the long bones, by which the lever motion is possible. A muscle originating on one bone and terminating on another either moves both bones toward each other or, if one attachment is fixed, the movable is drawn toward the fixed part.

We may class muscles as striated or voluntary and unstriated or involuntary. A third class, mixed, is represented by the heart muscle. The striated is represented by the skeletal muscles, and the unstriated by the thin muscular layers that form part of the wall of the stomach, intestines, bladder and other hollow organs.

RHEUMATISM.—This is an inflammation of the tissues that form the locomotory apparatus. The effect of cold on the muscles and tendons is an important factor in its production. It differs from other inflammations by shifting from one part to another. It is termed *muscular rheumatism* when it affects the muscles, tendons and fascia, and *articular rheumatism* when it involves the articulations. A second classification, *acute* and *chronic*, depends on the character of the inflammation. The muscular form is common in horses, dogs and hogs, while the articular form more commonly affects cattle.

The following causes may be considered. Animals that are exposed to cold, wet, changeable weather, or kept in cold, damp, draughty quarters frequently suffer from rheumatism. Under such conditions it is very probable that imperfect metabolism of body tissue occurs, and certain toxic products that are capable of irritating the muscles and articulations form. Clinical symptoms, and the presence of bacteria in the inflamed tissue indicate that bacteria and their toxins play an important part in the development of articular rheumatism. Heredity is said to be an important predisposing factor. One attack always predisposes the animal to a second.

The symptoms vary according to the severity of the attack. Local rheumatism is not accompanied by serious symptoms. The regions most commonly involved in local, muscular rheumatism are the shoulder, neck and back. The joints affected in the articular form are the knee, fetlock, hip, elbow and shoulder. The attack is usually sudden and accompanied by fever, more or less loss of appetite and soreness. Loss of control over the movement of the hind parts or walking on the knees may occur in the smaller animals. The larger animals show a slight or severe lameness. The affected muscle or articulation may be swollen, hot and tender. Pressing on the part with the hand or forcing the animal to move about may cause severe pain. Weakness and emaciation may occur in generalized and articular rheumatism, especially if suppuration takes place in the affected joint.

The prognosis is more favorable in muscular rheumatism than in the articular form. Both forms may become chronic. It is frequently advisable to destroy animals suffering from the articular form because of their emaciated, weakened condition and the deformed condition of the joints.

The preventive treatment consists in avoiding conditions favorable to the production of rheumatism. In ventilating the stable we should avoid draughts. Practical experience indicates that allowing a horse to stand in a draught after it has been warmed up by exercise is a very common source of muscular rheumatism and is especially to be avoided. Young hogs and sows that are thin are very prone to rheumatism when given wet, draughty sleeping quarters. Houses having dirt or loose board floors are very often draughty. Concrete floors when wet and not properly bedded with straw are objectionable. Although we do not fully understand the causative factors, we can take advantage of the knowledge we have gained from practical experience, and avoid keeping animals under conditions that are favorable for the production of the disease. It is almost useless to treat rheumatism unless the conditions under which it occurred are corrected.

The treatment is both local and internal. The local treatment consists in applying a mild liniment to the part, together with massage. If the part is tender and painful, hot applications may be used. Spirits of camphor ten parts and turpentine two parts, applied daily, are useful in relieving the soreness of rheumatic muscles. Salicylate of soda two ounces, fluid extract of gentian one ounce, and sufficient water to make an eight-ounce mixture may be given internally three times daily after feeding. Of the above mixture horses and cattle may be given one-half ounce and sheep and swine from one to two drachms. The treatment should be continued for a period of from eight to ten days or longer. It may be repeated in from one to two weeks.

Iodide of potassium is very useful in the treatment of chronic articular rheumatism. A very light diet should be fed and the animal given as complete rest as possible. An occasional physic should be given.

AZOTURIA, HAEMOGLOBINURIA.—This is a disease of solipeds affecting the muscles of the quarters. The affected muscles become swollen, hard and paralyzed. The disease follows a short rest, and rarely occurs when the animal is running in pasture or idle for a long period. Animals that are fat or rapidly putting on fat are predisposed to it. Animals that have had one attack are predisposed to a second.

The cause of this disease is not positively known. The German veterinarians attribute it to irritation of the muscles by cold, and classify azoturia as a rheumatic disorder. The conditions preceding the attack are not in favor of this theory, and cold can not be considered an important causative factor. The most acceptable is the auto-poisoning theory advanced by Dr. Law.

Azoturia is common in the country where feed is abundant and wrong methods of feeding horses are commonly practised. It is a very common practice to feed horses accustomed to hard work the same ration when idle for a few days as when working. The blood of horses cared for in this way may become abnormally rich in albuminoids. The suddenness of the attack, occurring shortly after the animal is given exercise, indicates auto-poisoning. This may be due to the blood in the portal vessels and the liver capillaries, charged with nutritious and waste products from the overfed animal's intestines, being suddenly thrown into the general circulation by a more active circulation of the blood brought on by exercise.

The symptoms of disease are manifested shortly after the animal is moved out of the stall and given exercise. When the animal is first exercised it is usually in high spirits. After travelling a short distance it is noticed to sweat more freely than ordinarily, breathe rapidly, lag and go lame, usually in the hind limbs. It trembles, shows evidence of suffering severe pain by turning its head and looking around toward the flanks, knuckles over in the hind pasterns, and may fall down and be unable to get up. The affected muscles appear to be swollen and feel unusually firm when pressed upon with the hand. If the horse does not go down recovery may occur within a few hours, and we are able to move the horse to the stable. Dark brown urine may be passed. At other times, the animal lies in a natural position, possesses a good appetite, but can not stand. In the severe form, it is restless and shows marked nervous symptoms.

The prognosis is unfavorable in the severe form. When nervous symptoms are absent recovery usually occurs in from two to ten days. Complications are common. More or less atrophy of the muscles of the quarters may result .

The preventive treatment consists in avoiding conditions that may favor the production of the disease. More attention should be given the feeding and care of work animals. If it is not possible to permit horses that are worked to exercise in a lot or pasture when idle, the ration should be reduced and roots, chopped, or soft feed given.

Careful nursing is an important part of the *treatment*. As soon as the horse shows evidence of an attack, it should be stopped and allowed to stand until sufficiently recovered to be moved. If paralysis occurs, we should make it as comfortable as possible and arrange to move it to a comfortable, warm, well-bedded stall. It may be advisable to place the animal in slings. This is not advisable in the serious form of the disease because of the extent of the paralysis and the nervous symptoms. A very light diet, bran mashes, chopped hay or green feed, should be fed during the convalescent period and for several days after complete recovery has occurred.

The following lines of *medicinal treatment* may be recommended. We should endeavor to stimulate the elimination of the waste products from the body by way of the kidneys, intestines and skin. This may be accomplished by administering saline cathartics, covering the body with blankets, encouraging the animal to drink plenty of water and feeding soft feeds. Glauber's salts may be given as a drench, or eserine may be given hypodermically. Sedatives such as chloral hydrate may be used to quiet the animal.

QUESTIONS

1. Give a general description of the locomotory apparatus.

2. Give the causes of rheumatism; describe the treatment.

3. What is azoturia? Give the cause of this disease.

CHAPTER XIV
STRUCTURE OF THE LIMBS OF THE HORSE

GENERAL DISCUSSION.—Each limb is formed by a column of bones that rest upon one another, forming more or less open angles. The bones of the column meet and form articulations that are held together by ligaments, and attached to their faces, borders and extremities are muscles and tendons. In the superior portion of the limb the muscles are heavy, tapering inferiorly, and terminating in the region of the foot in long tendons. Each limb is divided into four regions. The regions of the *fore-limb* are the shoulder, arm, forearm and forefoot. In the *hind limb* are the regions of the pelvis, haunch, thigh, leg and hind-foot. The feet in turn are divided into three sub-regions each. The *forefoot* is formed by the knee, cannon and toe, and the *hindfoot* by the hock, cannon and toe.

THE SHOULDER BONE OR SCAPULA is flat and triangular in shape. It is attached to the trunk by heavy muscles, one of which, together with its fellow on the opposite side, may be compared to a great, muscular sling that supports about two-thirds of the body weight. Attached to the internal and external faces of the scapula are heavy muscles that pass over the shoulder-joint, and become attached to the arm bone through the insertion of their muscular fibres or by a short tendon.

THE ARMBONE OR HUMERUS belongs to the class of long bones. Its superior extremity forms a flattened head that fits rather imperfectly into a shallow cavity in the humeral angle of the scapula. The inferior extremity resembles a portion of a cylinder in shape, and fits into shallow depressions in the superior extremity of the principal bone of the forearm. The muscles here are divided into two regions, anterior and posterior brachial. The most of these muscles originate on the posterior border and inferior extremity of the shoulder bone, and terminate inferiorly on the superior extremities of the principal and second or rudimentary bone of the forearm. The posterior brachial muscles are heavy and powerful. They are sometimes termed elbow muscles, because they are attached to the point of the elbow.

THE REGION OF THE FOREARM is formed by two bones, the *radius* and *ulna*. The radius is the principal bone and is classed among the long bones. The ulna is an elongated flat bone. It is attached to the external portion of the posterior face of the radius and extends above the superior extremity of this bone to form the point of the elbow. The radius articulates with the upper row of knee bones. The muscles of this region, the antibrachial, are divided into two sub-regions, anterior and posterior. They originate superiorly from the lower extremity of the arm bone and the superior extremities of the bones of the forearm, and terminate toward the lower extremity of the region in tendons that become attached to the bones of the knee, cannon and digit.

THE KNEE OR CARPAL region is formed by seven short bones that are arranged in two rows. They form a series of articulations. These are the articulations between the two rows, between the bones of each row, and between the upper and lower rows and the neighboring regions. Nearly all the motion takes place in the articulation between the upper row and the principal bone of the forearm.

THE CANNON OR METACARPAL region is formed by three bones. These are the principal metacarpal or cannon bone, and the rudimentary metacarpal or splint bones. The latter are attached to the margins of the posterior face of the cannon bone. The superior extremities of these bones articulate with the lower row of carpal bones. The convex extremity of the cannon bone meets shallow depressions in the superior extremity of the first digital bone. This is termed the fetlock joint. The anterior and posterior faces of this region are travelled by the long tendons belonging to the extensor and flexor muscles of the digit.

THE DIGIT OR TOE is formed by six bones, three of which are termed accessory or sesamoids. The digital bones may be given numerical names.

THE APPROXIMAL OR THIRD DIGITAL BONE is the shortest long bone in the body. The two shallow articular cavities belonging to the superior extremity are completed posteriorly by the two sesamoid bones. The inferior extremity is smaller than the superior and resembles the inferior extremity of the cannon bone in shape, excepting that it shows a middle groove. The anterior and posterior faces are travelled by the tendons of the digital muscles.

THE MIDDLE OR SECOND DIGITAL BONE is quite short. It articulates superiorly with the first, and inferiorly with the third bone of the digit. The superior face shows two shallow cavities, and the inferior two convex surfaces separated by a median groove. The latter face articulates with the third and navicular bones. The popular name for this articulation is the coffin joint.

THE THIRD OR DISTAL DIGITAL BONE may be compared to a cone that has been cut away posteriorly, obliquely downwards and backwards. The superior face shows two shallow cavities that are completed posteriorly by the superior face of the coffin or navicular bone. The anterior face is convex and cribbled by openings, and the inferior face is concave, forming the sole. Tendons belonging to the digital muscles terminate on the summit and inferior face of this bone.

THE PELVIS OR HAUNCH is formed by a single bone, the *coxa* that in the foetus may be divided into three bones. These are the *ilium, pubis* and *ischium*. It belongs to the class of flat bones. Anteriorly it is flattened from before to behind and directed inward and upward. The external angle is rugged and is generally termed the angle of the haunch. The internal face of the opposite angle articulates with the sacrum, to which it is firmly attached by ligaments. The middle portion is constricted and forms a neck. The inferior or posterior portion is flattened from above to below, and directed inward to meet the border of the opposite bone. Just below the neck and externally, there is a cup-shaped cavity into which the head of the thigh bone fits. The two coxa, together with the sacral ligaments (sacrum) and the muscles of the quarter, enclose the pelvic cavity.

THE REGION OF THE THIGH is formed by the *femur*, the largest long bone in the body. The superior extremity is formed by a rugged eminence, to which the heavy muscles of the quarter are attached, and by an articular head. The inferior extremity is formed by two convex articular surfaces that are separated by a deep notch, and a third pulley-like articular surface, with which the patella or knee-cap articulates. The pair of condyles articulates with the superior extremity of the leg bone. The thigh or femoral region is heavily muscled.

THE LEG is formed by three bones. The patella, a short bone, has already been mentioned as articulating with the thigh bone. The tibia and fibula are the other two bones in the region.

THE TIBIA belongs to the class of long bones and the fibula is quite rudimentary, being represented by a stylet-shaped bone that lies posterior to, and along the outer border of the tibia. The superior extremity of the tibia shows a central spine margined laterally by rather plain articular faces. It articulates with the thigh bone. The muscles of this region are divided into two sub-regions, *anterior* and *posterior* tibial. The muscles originate from the lower extremity of the femur and the two bones in this region, and terminate inferiorly in tendons that are attached to the bones of the hock, cannon and digit.

THE HOCK OR TARSAL region is formed by six bones. They are described as forming two rows. In the upper row there are two bones and in the lower four. They form a series of articulations, the same as the bones of the knee. Practically all of the movement occurs in the articulation between one of the large bones in the upper row and the lower extremity of the tibia. It may be mentioned here that this is the most perfect hinge-joint in the body. A very large tendon is attached to the summit of the hock. Other tendons cross and become attached to the hock bones.

The regions of the *hind cannon* and digit are practically the same as the corresponding regions of the forefoot.

QUESTIONS

1. Name the different bones of the fore-limb; hind limb.

2. Describe the regions of the shoulder, arm and forearm.

3. Describe the region of the forefoot.

4. Describe the regions of the haunch, thigh and leg.

5. Describe the region of the hindfoot.

CHAPTER XV
UNSOUNDNESSES AND BLEMISHES

GENERAL DISCUSSION.—The value of a horse depends largely on the condition of the limbs and their ability to do the work for which they are intended. This fact is frequently overlooked by experienced horsemen, who give attention to general conformation and action rather than to soundness of limb.

Diseases affecting the limbs may be classed as *unsoundnesses* and *blemishes*. This classification is based on the degree to which the disease interferes or may interfere with the work that the animal is called on to perform. Unsoundnesses interfere with the use of the part or the use of the animal for a certain work; blemishes do not. Such a basis for the classification of diseases does not enable us to place certain diseased conditions of the limbs in the unsound, or the blemish class at all times. A curb may, if it produces lameness, be classed as an unsoundness. If it does not cause the animal to go lame, and the enlargement on the posterior border of the hock is small, it is classed as a blemish. A high splint may place the animal in the unsound class, but usually a low splint is not considered a serious blemish. This classification is based to a certain extent on the relative economic importance of the disease, or the influence that the disease may have on the value of the animal, as well as any interference with the animal's ability to work.

RECOGNITION OF THE DISEASE.—The seat of the disease may be in a muscle, tendon, bone or ligament. The general symptom manifested is lameness or pain. The local symptoms are heat, pain, swelling and bony enlargements. The degree of lameness and the character of the local lesions vary greatly in the different cases. When the animal shows a slight lameness and there is little evidence of any local symptom, it requires the services of a skilled and experienced person to locate the diseased part. When the part shows local lesions of disease and the lameness is characteristic, diagnosis is not difficult.

THE EXAMINATION should be made while the animal is at rest; while standing in the stall and on level ground; when moved at a walk, or a slow trot on soft ground, or a hard roadway; and when moved out after resting a few hours. While examining the animal under the different conditions mentioned, the examiner must be careful and not pass over any part of a limb without determining whether it is normal or not. He should note any abnormal position that the animal may take while standing at rest. Every movement should be watched closely, as the manner of favoring the part may characterize the lameness. Negative symptoms of lameness in a part may at times prove as valuable in forming a diagnosis as positive symptoms.

The resting of either of the front feet, when the horse is standing at ease, indicates that there is some soreness in the rested limb. *Pointing* or placing one or both feet well in front of the line of support, when the animal is standing, usually indicates a diseased condition of the feet. It is natural for a horse that is standing in a stall to rest the hindfeet alternately. When the hindfoot is rested because of a soreness in some portion of the limb, it may be flexed or extended, the weight rested on the toe, and the foot flexed and bearing practically no weight. In serious inflammation of the front feet, both feet may be placed well in front of the normal position, and the hindfeet well under the body.

WHEN EXAMINING A HORSE, the blanket or harness should be removed. The horse should have on an open bridle or halter, and the attendant should give it as much freedom of the head as possible. The examiner should examine each limb carefully and note any symptom of disease that may be present. The attendant should walk the animal straight away from the person making the examination, toward, and past him, so that the animal's movements can be observed from both sides, from behind and in front. This examination should be repeated with the horse at a slow trot.

The character of the lameness may enable us to locate the seat of the disease. We must first determine in which limb the animal is lame. This part of the diagnosis is not difficult. The pain suffered every time weight is thrown on the diseased limb causes the horse to step quickly and shift as much of the body weight as possible on the well foot. The foot of the lame limb is jerked up rather quickly after weight is thrown on it. This favoring of the part varies in the different diseases. When the foot of the sound limb comes to the ground, more weight than common is placed on it. If the seat of the lameness is in a front limb, there is a decided nodding or movement of the head downward when the weight is placed on the well foot. If both forefeet are diseased, the animal steps shorter and more quickly than common. Lameness in a hind limb is characterized by more or less dropping of the quarter of the well limb when weight is thrown on it, and sometimes by a "hitch" or elevation of the quarter of the diseased limb when it is carried forward.

Unless there are *local symptoms of disease* present, it may be quite difficult to locate the seat of lameness. Sometimes local symptoms are misleading. After the lameness has been located in a certain limb, its movement must be carefully noted in order to detect the part favored. If the lameness is not characteristic enough to enable the examiner to locate the seat of it, it is then necessary to put the animal through some movement that may emphasize the soreness in the part. The animal may show a certain reluctance to throw weight on the limb when turned to the right or left. Moving the horse in a small circle with the lame limb on the outside may cause the animal to use the muscles of the shoulder more freely, and emphasize any soreness that may be present. If the lame limb is on the inside, soreness anywhere in the foot may be increased, because of the extra weight thrown on this portion of the limb. Moving the animal over a hard driveway may increase the pain resulting from an inflammation of the feet. Causing the animal to trot on soft ground, step over high objects, flexing, extending, abducting and adducting the part may enable the examiner to locate the exact group of shoulder or arm muscles involved by the disease.

IN EXAMINING THE FEET it may be necessary to remove the shoes and practise percussion and pressure over the region of the sole. In some forms of lameness it may be necessary to destroy the sensation in the foot by injecting cocaine along the course of the nerves that supply the foot before arriving at a definite diagnosis.

QUESTIONS

1. Define the term unsoundness and give an example.

2. Define the term blemish and give an example.

3. Give the general method of examining a horse for soundness.

CHAPTER XVI
DISEASES OF THE FORE-LIMB

SPRAINS AND INJURIES IN THE REGION OF THE SHOULDER.— Sprains and injuries of the structures in the shoulder region are more common in horses that are called on to do heavy work than among driving horses.

The following *causes* may be mentioned: Ill-fitting collars, pulling heavy loads over uneven streets or soft ground, where the footing is not secure, and slipping are common causes. Young horses that do not know how to pull, or horses that are tired out by hard work, are predisposed to muscular strain, and are apt to suffer injury if forced to do heavy work. Sore shoulders, or an ignorant driver, may cause the animal to pull awkwardly and throw more strain on certain groups of muscles than they can stand. Rheumatism frequently causes shoulder lameness. The muscle usually affected by rheumatism is the large muscle extending from the region of the point of the shoulder to the summit of the head.

The symptoms of shoulder lameness vary in the different cases. The horse may walk without going lame, but when made to trot lameness is quite noticeable. The animal may point with the foot of the diseased limb, holding it forward, but squarely on the floor. In severe strain, little weight is thrown on the limb and the lameness is marked . In "shoulder slip" the head of the arm bone pushes outward every time the animal throws weight on the limb. This luxation can be noticed best when standing in front of the animal. Marked atrophy of the external shoulder muscles may occur. Such atrophy may appear and disappear quickly, and may result from an injury to the nerve supply of the muscle as well as from favoring the part. Atrophy of the shoulder may occur if the animal is lame in other regions of the limb, especially the feet. The outcome of shoulder lameness is favorable if the disease causing it is given prompt treatment.

Rest is a very important part of the *treatment*. It may be advisable to restrict the horse's movements by placing it in a single stall, and tying the animal so that it can not lie down. This should be continued for at least one week. If the horse is restless, it should be given a box-stall or turned out in a small lot alone. It should be watered and fed in the quarters where confined. The *local treatment* consists in applying mild liniments or blisters to the shoulder. It is not advisable, however, to apply a blister if the muscles feel hot and tender.

CAPPED ELBOW, "SHOE-BOIL."—Capped elbow is an inflammation of the bursa at the posterior surface of the elbow . The swelling that results is usually sharply defined. It may feel abnormally warm and doughy, and it may be painful. Later, the enlargement may be well defined and hard. Sometimes the skin is indurated and lies in folds, or the shoe-boil shows abrasions on its surface and fistulous openings leading from abscess centres. The cystic or soft tumor is a common form. Such an enlargement fluctuates on pressure, and when opened, a blood-stained fluid escapes. All forms of capped elbow tend to become chronic.

The treatment is both preventive and local. As capped elbow is caused by bruising the part with the hoof or heel of the shoe, the preventive treatment consists in hindering the animal from taking a position that may favor injury to the part. Confining the animal in a small stall and tying it with too short a halter strap favors a sternal position when lying down. A roomy stall that permits the animal to stretch or change position is an important preventive measure. Shoes that project beyond the quarters should be avoided. The elbow may be protected by placing a thick pad over the heels when the animal is in the stable.

Local treatment varies according to the character of the enlargement. Soft, doughy swellings may be treated by application of cold, iodine and blisters. The cystic form of tumor must be opened, the fluid removed and the lining membrane destroyed by the injection of tincture of iodine. Hard, indurated shoe-boils may be treated by completely removing the diseased tissue. The surgical treatment of capped elbow requires the service of an experienced veterinarian. His efforts may prove a complete failure, unless the irritation to the part by the shoe or hoof is prevented.

INJURIES TO THE KNEE (BROKEN KNEE).—Horses frequently fall and bruise or lacerate the knee when moving at trot or canter. The injury varies according to the force of the fall, and the character of the road that the animal is travelling over. Some individuals are more liable to suffer from this class of injuries than others. Horses that are weak-kneed because of poor conformation, or knee-sprung, are inclined to stumble. Careless driving, especially if the animal is tired, predisposes it to this class of injury. Because of the predisposition toward stumbling on the part of some horses, scars on the front of the knee are termed broken knee, and the animal is considered unsound.

The symptoms vary with the extent of the injury. Slight bruises or abrasions result in local swelling and soreness that disappear within a few days. Laceration of skin interferes with the movement of the knee and the animal may be quite lame. The part becomes swollen and painful. In injuries involving the sheaths of the tendons and the synovial membrane, the pain is severe and the accompanying inflammation may take on a serious form.

The preventive treatment should not be overlooked. Horses should be trained to carry the head at a proper height when moving. The driver should handle the reins properly and keep his attention on the horse or horses that he is driving. Superficial bruises require no special treatment other than rest. Laceration of the skin and underlying tissue requires complete rest and careful removal of any particles, of dirt and gravel that may be present in the wound. Shreds of tissue that may take no part in the healing should be cut away. The hair in the region of the wound should be trimmed short. Careful and repeated dressings with antiseptics are necessary until the inflammation has largely disappeared and healing is rapidly taking place. It may be advisable to tie the horse in the stall so that it can not lie down.

DISTENDED SYNOVIAL SACS, JOINT SHEATHS AND BURSAE, "GALLS."—Soft enlargements may occur in the region of the knee and fetlock. They are commonly termed "galls," "wind-galls," or "road-puffs." They are usually due to the sheaths surrounding the tendons becoming distended with synovia. "Galls" are caused by strains, direct injury to the part and severe, continuous work. Certain individuals may develop this class of blemishes without being subject to any unusual conditions. This condition is seldom accompanied by lameness.

The treatment may vary in the different cases. If the distended sheath, or bursal enlargement, is caused by a direct injury or strain, cold bandages should be applied and the part given as complete rest as possible. "Wind-galls" may be removed by a surgical operation. It is not advisable to attempt the removal of "road-puffs." Rest, stimulating leg washes and bandages may temporarily remove the latter.

SPRUNG KNEES (BUCK KNEES).—This condition of the knee is characterized by the partly flexed condition of the region. It is best observed by standing to one side of the horse . Instead of the forearm and cannon regions appearing perpendicular or in line, they are directed forward. This condition may exist in varying degrees. Some individuals show it to a slight degree, the condition being accompanied by a weakness or shakiness of the knee when standing at rest. Sometimes, but one knee is involved.

The causes of this unsoundness are hereditary and accidental. Weak knees due to faulty conformation seldom escape becoming sprung in animals that are given hard work. Severe and continuous driving is a common factor in the production of this condition. Strains of the flexor muscles of the region may cause it. The retraction of the flexor muscles and their tendons and the aponeurosis of the antibrachial region occurs in this disorder and prevents the animal from extending the knee.

The region is greatly weakened by this condition and the animal may be unfitted for active work. For this reason the value of the animal is greatly diminished.

Treatment is unsatisfactory. The preventive treatment consists in not breeding animals that have poorly conformed knees and using the proper judgment in working young horses and when driving or riding horses. Certain cases may be greatly benefited by sectioning the tendons of the external and middle flexors of the metacarpi. To insure a successful outcome in any case that is operated on, a long period of rest is required.

SPLINTS.—A splint is a bony enlargement situated along the line of articulation between the splint and cannon bones . This blemish is due to an inflammation of the periosteum. It is a very common blemish and is generally located along the splint bones of the forefeet, especially the internal ones.

Splints are caused by strains and rupture of the ligament that binds the splint bone to the cannon bone. The result is an inflammation of the periosteum. Slipping, or an unbalanced condition of the foot, may cause this injury by distributing the weight unequally on the splint bones. Faulty action and bad shoeing may cause the horse to strike and bruise the region.

Symptoms of lameness are not always present. A high splint involving the articulation between the lower row of carpal, splint and cannon bones may be considered an unsoundness, because of the persistent character of the lameness. The animal may show little or no lameness when walked, but if moved at a trot, especially over a hard roadway, it may show marked lameness. The local inflammation is characterized by a small swelling lying along the splint bone, that feels hot and may pit on pressure. After a time the inflammation disappears and is replaced by a hard, bony enlargement. When this occurs the lameness disappears.

The preventive treatment consists in keeping the feet of young horses in proper balance by frequent trimming and proper shoeing. This attention is very necessary in young colts that are running in pasture. It is very advisable to rest the animal during the period of inflammation. Cold bandages should be applied. As soon as the inflammation has subsided mild counterirritants and absorbents may be used. In. case the lameness persists, more severe counterirritation is indicated.

INFLAMMATION OF THE FLEXOR TENDONS OF THE DIGIT.—
The large tendons posterior to the foot and the suspensory ligament
that separates them from the cannon bone frequently become inflamed.
Sometimes complete rupture of one or more of these structures occurs.
The lighter breeds of horses are the most frequent sufferers. Because of the
greater strain thrown on the tendons of the forefeet, inflammation of these
tendons is far more common than it is in the hindfoot. Diseased conditions
of the hind tendons are usually due to other causes than strain.

The following *predisposing* and *accidental causes* should be considered:
Weak flexor tendons and heavy bodies predispose animals to inflammation
of the tendons and suspensory ligament; quality, not size, is the factor to
consider when judging the strength of a tendon; long, slender pasterns
increase the strain on these structures, and this mechanical strain is further
increased by low heels and long toes; the character of the work and the
condition of the road that the animal travels over are important factors to
consider; trotting and running horses more often suffer from injuries to
tendons and ligaments than draft horses; travelling at a high rate of speed
over an uneven road, slipping and catching the foot in a rut or car track, are
common causes; bruises and wounds may result in the tendons becoming
inflamed; inflammation of the tendinous sheaths and the tendons as well
sometimes occurs in influenza.

Lameness is a prominent *symptom*. The pastern is held in a more upright
position than normal. When the animal is standing, the foot is rested on
the toe, and it may take advantage of any uneven place on which to rest
the heel. In severe strains the local symptoms are quite prominent. The
tendons may be hot and swollen. Pressure may cause the animal pain. In
chronic tendinitis the tendon may be thickened and rough or knotty. Pain is
not a prominent symptom in this class of cases. Shortening of the inflamed
tendon may occur, causing the animal to knuckle over. Rupture of one or
more of the tendons and the suspensory ligament can be recognized by the
abnormal extension of the pastern. If the ruptured tendon heals, it always
results in a thickening at the point of the rupture that gives the tendons a
bowed appearance. This is termed bowed-tendon.

The lameness resulting from an inflammation of tendons resembles
that resulting from strains and injuries to the fetlock joint, especially in the
region of the sesamoid bones.

INFLAMMATION OF THE SESAMOID BONES differs slightly from
the former. Pressure over the posterior region of the fetlock may cause the
animal pain. The lameness shows a tendency to disappear with rest and
reappear when the animal is again worked. Lameness is most prominent

in some cases when the animal is first moved out. There may be a lack of local symptoms, such as heat and swelling. It is not uncommon for a bony enlargement to form on the sesamoid bone after a few months or a year.

The following *treatment* is recommended. Horses that have a poor quality of tendon and weak fetlocks and pasterns should not be used for breeding purposes. Careful driving would prevent a large percentage of injuries to tendons. The most important treatment for all injuries due to strains is rest. In all cases of severe strain to the structures in this region, it is very advisable to apply a plaster bandage. This should be left on for at least two weeks. When the acute inflammation has subsided, counterirritants may be applied. Either cold or hot applications are recommended. Cold applications are to be preferred at the beginning of the inflammation. Covering the tendons with a cold bandage, or with a heavy layer of antiphlogistin, is recommended. The horse should not be worked until after the tendons have had an opportunity to completely recover from the inflammation.

CONTRACTED TENDONS, KNUCKLING-OVER.—New-born foals are sometimes unable to stand on their front feet because of the excessive knuckling-over. The colt may walk on the front of the pastern and fetlock. This sometimes results in severe injury to the skin and the underlying tissues.

Knuckling-over in the mature horse is not always due to contracted tendons. It may occur as a symptom of inflammation of the flexor tendons, ligaments of the fetlock joint and the articulation as well. It may be noticed in animals that have ring-bone, or coffin-joint lameness.

The most *common cause* for this unsoundness is inflammation of the muscles and tendons of the flexors of the digit. As a result of long standing or severe inflammation, shortening of these structures occurs in consequence of the contraction of the inflammatory or cicatricial tissue. Knuckling-over in the newborn colt is commonly caused by a weakness or lack of innervation of the extensor muscle of the digit. Judging from the quick recovery that usually occurs, other causes for this condition are seldom present.

The treatment recommended for the new-born colt is supporting the fetlock with a light plaster bandage. This should be applied very soon after birth in order to prevent bruising of the fetlock. A light cheese-cloth bandage should be applied to the limb from the hoof to the knee. The colt is laid on its side, the toe extended as much as possible, and the plaster bandage applied. This should be removed in about one week and fresh bandages applied. In about two weeks the young animal is usually able to walk on the toe. As soon as it is able to do this a bandage is unnecessary. It is not advisable to turn the colt outside if there is any chance for the bandages to become wet.

Knuckling-over due to faulty conformation is difficult to correct. Light work and careful shoeing are the most valuable preventive measures in young horses. Sprains and injuries to the region of the fetlock should receive the necessary treatment. The treatment for contracted tendon is largely surgical and consists in sectioning it.

INJURIES CAUSED BY INTERFERING.—Horses that have faulty action may strike the opposite fetlock with the moving foot, the inside of the opposite limb in the region of the knee, and the quarters of the front foot with the shoe of the hindfoot. It is very common for horses to "brush" the inside of the hind fetlock with the opposite foot when trotting, especially if tired. Interfering in the front feet is less common. Striking the inside of the region of the knee with the opposite foot or *"speedy cutting"* occurs in driving and speed horses. Both of the latter forms of interfering may be considered unsoundnesses.

The most *common cause* of interfering is faulty conformation, such as narrowness of the chest or pelvis, faulty conformation of the limbs and irregularity in the action of the joints. Shoeing and the condition of the feet are also important factors. Animals that have a narrow chest or pelvis interfere because the legs are placed too closely together. Turning in of the knees or "knock-kneed," winging in or out of the feet, or any other defective conformation of the limbs that tends to prevent the animal from moving the feet in line, lead to serious interfering. A wide-spreading hoof, an unbalanced condition of the foot and improper fitting of the shoes are common causes for interfering in horses that would otherwise move the feet in line. Debility from disease and overwork may cause the animal to interfere temporarily. An unbalanced gait and shortness of the body are the common causes for injuries to the quarters.

All degrees of injury to the part struck by the shoe or wall of the foot may be noted. Horses that interfere lightly, wear the hair off and produce slight abrasions of the skin on the inside of the fetlock. Sometimes the skin is bruised, inflamed or scarred. Injuries to the inside of the knee and quarter are the most serious. Lameness, inflammation of the periosteum and bony enlargement may result from "speedy cutting." Deep wounds in the region of the heel or quarter may occur when a horse strikes this part with the shoe of the hindfoot in moving at a high rate of speed.

The treatment is largely preventive. No doubt many cases of interfering could be prevented by careful training and balancing of the foot when the animal is growing and developing. The feet of colts should be trimmed every three or four weeks. Interfering in the hindfeet may be stopped by noting the character of the animal's gait and the portion of the wall that strikes the part, and by practising intelligent methods of shoeing. Slight injuries should be treated by the application of antiseptic powders. The treatment for injuries to the periosteum is the same as that recommended for splints. As a last resort boots and button rings may be used for the purpose of preventing serious injury to that part which is struck by the foot.

RING-BONE.—Chronic inflammation of the articulation between the first and second bones of the digit is termed ring-bone . Not all ring-bones involve the articular surfaces. The periarticular, or false ring-bone, is a chronic inflammation of the bone near the articular surface. The bony enlargement varies in size. It may form a ring encircling the part, or it may be limited to the lateral surface of the joint. The bony enlargement may be so small as to be detected only by a careful examination. Ring-bone may occur on any of the feet, but it is said to be more common in the front feet.

The predisposing cause of ring-bone is faulty conformation. Long, weak pasterns that are predisposed to strains, upright pasterns, especially if small, and exposed to concussion and jarring, and crooked feet that distribute the weight on the part irregularly are important factors in the production of ring-bones. The *external causes* are sprains or any injury to the region. *Lameness* is nearly always present. The degree of lameness varies, and does not depend altogether on the size of the bony enlargement. Large ring-bones interfere with the movement of the tendon. Lameness is most pronounced when weight is thrown on one foot, the later phase of the step being shortened and the pastern more upright. Some cases improve with rest, but the lameness returns when the animal is given hard work.

The preventive treatment consists in giving the necessary attention to the feet of young animals, by trimming the wall frequently and keeping the feet in balance and the careful selection of breeding stock. Resting the animal, keeping the foot that has the ring-bone on it in proper balance and counterirritation by means of blisters and cautery (searing) are important lines of treatment. Shortening the toe and raising the heel, if necessary, greatly relieves the lameness in some cases. Sectioning the sensory nerves that go to the part should not be practised, unless in exceptional cases.

QUESTIONS

1. Give the causes of shoulder lameness; give the treatment.

2. Describe capped elbow; give the treatment.

3. What is "broken knee"?

4. What are "wind-galls" and "road-puffs"?

5. Give the cause and treatment of sprung-knee.

6. Give the cause and treatment of splints.

7. What class of horses most commonly have strained tendons? Give the causes and treatment of this form of lameness.

8. Give the treatment of contracted tendons in the new-born colt.

9. Give the causes for interfering.

10. What are the different forms of ring-bone? Give the causes and treatment of ring-bone.

CHAPTER XVII
DISEASES OF THE FOOT

GENERAL DISCUSSION.—The foot of the horse as generally spoken of, includes the hoof and the structures that are enclosed by it . It may be divided into three parts, the insensitive and sensitive structures and the bony core. The *insensitive foot* or *hoof* is divided into wall, sole, frog and bars. The *sensitive foot* is divided into vascular tissue and elastic apparatus. The vascular tissue is in turn divided into coronary cushion, laminae and velvety tissue. The elastic apparatus is divided into plantar cushion and fibro-cartilages. The *bony core* is formed by the navicular and third digital bones. The hoof and vascular tissue in turn enclose the elastic apparatus and bony core.

THE WALL forms that portion of the hoof seen when the foot rests on the ground . It is covered by a thin layer of horny tissue, the *peripole,* that coats over the wall and assists in preventing its drying out. On lifting the foot and examining its inferior surface, it is noticed that the wall at the heels is inflected under the foot and in a forward direction. This portion of the wall is termed the *bars.* Within the bearing margin of the wall and in front of the bars is a thick, concave, horny plate that forms the *sole.* At the heels and between the bars is a wedge-shaped mass of rather soft horny tissue that projects forward into the sole. This is the *foot pad* or *horny frog.* It is divided into two lateral portions by a medium cleft.

THE CORONARY CUSHION projects into the upper border of the wall. It is covered with vascular papillae which secrete the horny fibres that form the wall. The *vascular laminae* are leaf-like projections, the sides of which are covered by secondary leaves. *Horny laminae,* arranged the same as vascular laminae, line the wall. These two structures are so firmly united that it is impossible to tear them apart without destroying the tissue. The *velvety tissue* covers all of the inferior surface of the foot, with the exception of the bars. As the name indicates, its surface is covered by vascular papillae that resemble the ply on velvet. It is firmly united to the horny sole which it secretes.

The lateral cartilages are attached to the posterior angles of the pedal bone. They are flattened from side to side, and the portion that projects

above the coronary cushion may be felt by pressing on the skin that covers it. The *plantar cushion* is a wedge-shaped piece of tissue formed by interlacing connective-tissue fibres and fat. It is limited on each side by the lateral cartilages. Its inferior face is moulded to the frog.

THE BONY CORE formed by the last bone of the digit and the coffin bone was described briefly with the other foot bones. A very important bursa, because it is so frequently inflamed in coffin-joint lameness, facilitates the gliding of the flexor tendon over the navicular bone before it becomes attached to the inferior face of the pedal or digital bone.

SIDE-BONES.—This is a chronic inflammation of the lateral cartilages of the foot that results in their ossification . This unsoundness is common in heavy horses, especially if worked on city streets. The inflammation affects the cartilages of the front feet, rarely those of the hindfeet.

The hereditary tendency toward the development of side-bones is an important predisposing factor. It is not uncommon to meet with this unsoundness in young horses that have never been worked. Low, weak heels, flat, spreading feet, or any other faulty conformation of the foot are predisposing factors.

The character of the work is an important *exciting cause.* Continuous work over paved streets, especially if the horse is shod with high-heeled shoes, increases the shock received by the elastic apparatus of the foot. This produces more or less irritation to the lateral cartilages, which may result in their complete ossification. Punctured wounds in the regions of the cartilage may cause it to become inflamed and changed to bone.

The following *symptoms* may be noted. Farm horses that have side-bones seldom show lameness. This is because they are worked on soft ground and not on a hard street or road. Driving and dray horses may step short with the front feet, or show a stilty action. This may disappear with exercise. The lameness is sometimes marked. The local diseased changes are the greatest help in the recognition of side-bones. Horses should not be passed as sound without making a careful examination of the lateral cartilages. This examination is made by pressure over the region of the cartilage with the thumb or fingers. This is for the purpose of testing its elasticity. If it feels rigid and rough, the cartilaginous tissue has been replaced by bony tissue, and the animal should be classed as unsound.

The treatment is largely preventive. Horses with side-bones should not be bred. It is not advisable to use horses with side-bones on the road or city streets. Shoeing with rubber pads may help in overcoming the concussion and relieve the lameness. Sectioning the sensory nerves going to this portion of the foot is advisable in driving horses. Rest and counterirritation relieve the lameness for a short time.

NAVICULAR DISEASE.—In navicular disease the bursa, flexor tendon, and navicular bone may become chronically inflamed. Because of the seat of the lameness, it is commonly known as "coffin-joint" lameness. This disease affects standard and thoroughbred horses more often than it does the coarser breeds. One or both front feet may be affected .

Hereditary causes are largely responsible for navicular disease. The tendency toward this disease probably depends on such peculiarities of conformation as narrow, weak, high heels, long pasterns and too long a toe. The character of the work is an important factor. Hurried, rapid movements throw considerable strain on the navicular region, increasing the danger from injury. This is, no doubt, one reason for "coffin-joint" lameness being more common in driving and speed horses than in slow-going work animals. Rheumatic inflammation, bad shoeing and punctured wounds in the region of the bursa many cause it.

The *first symptom* usually noted is a tendency to stumble. When standing in the stable, the animal "points" or rests the diseased foot. Sometimes it rests the heel of the lame foot on the wall of the opposite foot. If both feet are affected, the animal may rest them alternately, or take a position with both feet well in front of the normal position. The inflamed structures are so covered by other tissues that it is difficult to detect the local inflammation, or cause the animal to flinch by applying pressure over the region. As the disease becomes more advanced, the lameness becomes permanent. The limb is carried forward stiffly and rapidly and the animal stumbles when travelling over rough ground. In time, because of the little weight thrown on the posterior portion of the foot, the quarters may become higher, contracted and more upright and the frog smaller. If one foot is diseased, it becomes smaller than the opposite foot.

The following *preventive measures* may be recommended. We should not use animals having faulty conformation of the feet for breeding, because the offspring of such individuals have an inherent tendency toward navicular and other foot diseases. Animals that have "coffin-joint" lameness should be allowed to run in pasture as much as possible, because natural conditions help to keep down the inflammation and soreness and promote a more healthy condition of the foot. In shoeing the horse it is best to shorten the toe and raise the heel. It is advisable in the more favorable cases to cut the sensory nerves of the foot. This operation destroys the sensation in the foot, and should not be performed on feet with weak heels, or that are wide or spreading.

CONTRACTED QUARTERS.—This condition of the feet is characterized by the foot becoming narrow in its posterior portion. One or both of the quarters may be affected. It is principally observed in the forefeet.

The *causes* of contraction of the foot may be classed as *predisposing*, *secondary* and *exciting*. It may accompany chronic diseases of the foot, such as navicular disease and side-bones. Weak heels is the principal predisposing factor. Any condition that tends to prevent the hoof from taking up moisture, or causes it to lose moisture, may cause the horn to lose flexibility and contract. This is one of the reasons why horses that are worked continuously in cities, or used for driving, frequently develop contracted feet. Ill-fitting shoes, excessive rasping of the wall and bars, and allowing the shoes to stay on the foot for too long a time are responsible to a very large degree for this disorder of the foot .

The following *local symptoms* may occur: The wall of the foot at the quarters may appear drawn in at its superior or inferior portion. Sometimes one or both quarters are perpendicular, or nearly so. The foot then appears too narrow at the heel, too elongated and less rounded than normal. The changes in the appearance of the inferior surface of the hoof vary. The changes here may be so slight that they are not noticed. In well advanced and neglected cases the arch of the sole is increased, the frog is narrow and atrophied and the bars high and perpendicular. Corns may accompany the contraction. The foot may feel feverish. The animal may manifest the pain in the feet when standing at rest by pointing and changing their position. When lameness is present, it may resemble that occurring in inflammation of lateral cartilages and navicular disease.

Preventive treatment is of the greatest importance. This consists in giving the feet an opportunity to take up moisture when they are exposed to abnormal conditions and become feverish. Under such conditions, it is advisable to occasionally remove the shoes and turn the animal into a pasture or lot. It is best to do this in the fall or winter when the ground is wet. If this can not be practised, the shoes should be removed and a poultice of ground flaxseed and bran, equal parts, applied to the feet for a period of eight or ten hours, daily for a week or two. A plank trough six inches deep, two feet wide and as long as the stall is wide may be filled with a stiff clay, and the horse made to stand with its front feet in the clay bath for ten or twelve hours daily. When grooming the horse, the foot should be cleaned with a foot-hook and washed with clean water. Hoof ointments should be avoided so far as possible. The importance of fitting the shoe to the foot, avoiding the too free use of the rasp and hoof knife and resetting or changing the shoe when necessary can not be overestimated. Shoeing the animal with a special shoe is sometimes necessary. It is not advisable to attempt the forcible expansion of the quarters. Lowering the heels by careful trimming of the wall and sole and permitting frog pressure may be all the special attention required.

SAND-CRACK.—A fissure in the wall of the foot running in the same direction as the horny fibres, or a seam in the wall resulting from the healing of the fissure is termed sand-crack. The position and extent of the fissure or seam vary. It may involve the wall of the *toe* (toe-crack) or *quarter* (quarter-crack) . It is *superficial* or *deep*, according to the thickness of the wall involved; *complete* or *incomplete*, depending on whether it extends from the bearing margin of the wall to the coronary band or only a portion of the distance; *simple,* when the horny tissue only is involved; and *complicated,* when the sensitive tissue beneath becomes injured and inflamed. Cracks of long standing usually have thick, rough margins.

The causes of this unsoundness are poor quality of horn, improper care and injuries. Sand-cracks commonly occur in hoofs that are dry and brittle and have thin walls. In young horses incomplete cracks due to the wall becoming long and breaking off in large pieces are common. Unequal distribution of weight, the result of unskilled shoeing, or any other condition that may cause the foot to become unbalanced, using the foot rasp too freely, and such diseases as quittor, corns and contracted quarters subject the animal to this form of unsoundness. Any injury to the coronary cushion that secretes the fibres of the horny wall may result in either toe- or quarter-crack. Treads and barb-wire cuts are common injuries to the region of the coronet.

The preventive treatment consists in preserving a healthy condition of the horn by giving the foot the necessary care and attention in the way of proper trimming and shoeing, and providing it with the necessary moisture by means of foot-baths, wet clay and poultices. Quarter-cracks respond to treatment more quickly than toe-cracks. The treatment is practically the same for both. This consists in preventing motion in the margins of the fissure so far as possible.

The treatment for fissures in the region of the toe and quarter is as follows: The wall should be cut away along the margins of the crack until it is quite thin; and extra nail holes should be made in the shoe, and a nail driven into the bearing margin of the wall a little to each side of the fissure. The wall at the toe should be shortened and the toe of the shoe rolled if the animal's work permits the use of this kind of a shoe.

The margins of a quarter-crack and the wall just posterior and below it should be cut away until quite thin. The bearing margin should then be trimmed so that it does not rest on the shoe. A bar shoe that does not press on the frog may be used. Light blisters to the region of the coronet help in stimulating the growth of the horn. Rest is advisable.

CORNS.—This term is applied to injuries to the foot caused by bruises or continuous pressure to the posterior portion of the sole. This condition is common in the forefeet.

The predisposing causes are faulty conformation that favors pressure from the shoe on the sole between the bars and wall and weak heels. Corns are commonly met with in feet having contracted quarters. The principal *external causes* are wrong methods of shoeing and allowing the shoes to remain on the feet for too long a period.

A common symptom of corns is lameness. In order to relieve the pressure over the inflamed part, the animal stands with the foot slightly flexed at the fetlock. The lameness is not characteristic. It is only by a local examination of the foot, made by pressing on the sole or cutting away the horn, that we are able to form a positive diagnosis.

We describe the *diseased changes* by using the terms *dry, moist* and *suppurative corns*. In the *dry corn* we find the horn stained and infiltrated with blood. In the *moist corn* the hoof may be colored the same as in the former, but in addition there is a space between the vascular and horny tissue that is filled with a serous-like fluid. If this collection of fluid becomes infected with pus organisms and is changed to pus, it is then termed a *suppurative corn*. Sometimes the pus pushes its way upward and backward between the sensitive laminae and the wall, and makes its appearance at the margin of the coronary band in the region of the quarters or heels. This usually occurs when the tissues beneath the horny frog become bruised or the sensitive tissue pricked by a nail. It is commonly termed "gravelled." Pus rarely breaks through the thick horny tissue, but follows the wall and breaks through the skin where it meets with the least resistance. Corns may be considered a serious unsoundness in driving horses.

The treatment is largely preventive. Trimming the foot and fitting the shoe properly are important preventive measures. The practice of cutting away the bars and sole or "opening up the heels," as it is commonly termed, should be condemned. This method of trimming the foot instead of preventing corns is a very common factor in producing them. The shoe should not be too short or too narrow. It should follow the outline of the wall and rest evenly on its bearing margin. If this is practised, weakening the wall by cutting off that portion allowed to project beyond the shoe is unnecessary. Feet that have low heels and large, prominent frogs should be shod with shoes thick at the heels. The best line of treatment for a horse that is subject to corns is to remove the shoes and allow the animal to run in a pasture. If this is impossible, poulticing the feet or standing the animal in moist clay will help in relieving the soreness. Excessive cutting away of the horny sole is contra-indicated. Suppurative corns should be given proper drainage and treatment.

LAMINITIS, "FOUNDER." — This is an inflammation of the sensitive or vascular stricture of the foot. The inflammation may be acute, subacute or chronic. Stockmen frequently use a classification for laminitis based on the causes. Feed, road and water founder are common terms used in speaking of this disease. The inflammation is usually limited to the front feet.

The causes of laminitis are overfeeding, sudden changes in the feed, drinking a large quantity of water when the animal is overheated, overexertion, exhaustion and chilling of the body by standing the animal in a cold draft. It may be associated with such diseases as rheumatism, influenza and colic.

The symptoms vary in the different forms of the disease. Pain is the most characteristic symptom. The sensitive or vascular structure of the foot has an abundant supply of sensory nerves, and, as it is situated between the hoof and the bony core, the pressure and pain resulting from the inflammation are severe.

In the *acute form* general symptoms are manifested. The appetite is impaired, the body temperature elevated and the pulse beats and respirations quickened. If the inflammation is severe, the animal prefers to lie down. This is especially true if all four feet are inflamed. In most cases the horse stands with the forefeet well forward and the hind feet in front of their normal position and under the body. The affected feet are feverish and very sensitive to jarring or pressure. Moving about increases the pain in the feet, and it may be very difficult to make the animal step about the stall.

In the *subacute form* the symptoms are less severe. The irregularity in the gait is especially noticeable when the animal is turned quickly. The local symptoms are less marked than in the acute form and the general symptoms may be absent.

The chronic form is characterized by changes in the shape and appearance of the hoofs . The wall shows prominent ridges or rings, the toe may be concave, thick and long and the sole less arched than usual, or convex. The degree of lameness varies. It is more noticeable when the horse is moved over a hard roadway than if moved over soft ground. One attack of laminitis may predispose the animal to a second attack.

The prognosis depends on the character of the inflammation and the promptness and thoroughness of the treatment. Acute laminitis may respond to prompt, careful treatment in from ten to fourteen days. Subacute laminitis responds readily to treatment. The prognosis is least favorable in the chronic form.

The preventive treatment is very important. Dietetic causes are responsible for a large percentage of the cases of this disease. Horses that are accustomed to being fed and watered at irregular periods and after severe or unusual exercise seem to be able to stand this treatment better than animals that are more carefully cared for, but even this class of animals do not always escape injury. Stockmen should realize the danger of producing an inflammation of the feet by feeding grain and giving cold water to horses immediately after severe exercise. Overfeeding should also be avoided. Careful nursing may prevent the occurrence of laminitis as a complication of other diseases.

The treatment of the inflammation is as follows: The removal of the shoes and the necessary trimming of the foot should be practised early in the inflammation; the horse should be placed in a roomy box-stall that is well bedded with cut straw; during the cool weather it may be necessary to blanket the animal; if the weather is hot and the flies annoy the patient, the stall should be darkened; in serious cases, and when the animal is heavy, it may be advisable to use a sling; hot water fomentations are to be preferred; the patient may be stood in a tub of hot water or heavy woollen bandages that have been dipped in hot water and wrung, out as dry as possible may be applied to the feet; the temperature of the water should be no hotter than can be comfortably borne with the hands; the results of this treatment depend on the faithfulness with which it is carried out; a poultice of ground flaxseed should be applied to the foot at night, or during the interval between the foot-baths. This treatment may be continued until the acute inflammation has subsided.

If the animal is inclined to eat, it should be fed very little roughness and grain. Soft feeds are to be preferred, and one quart of linseed oil given as a physic. After a period of from ten days to three weeks, depending on the tenderness of the feet, the wall at the toe should be shortened, the sole trimmed if necessary, flat shoes rolled at the toe placed on the feet, and the animal allowed to exercise a short time each day in a lot or pasture. As the hoof grows rapidly, it is necessary to trim it carefully every three or four weeks and replace the shoes. The wall at the toe should be kept short, but excessive thinning of the sole should be avoided.

The same line of treatment as recommended for the horse may be used for laminitis in cattle. If marked diseased changes occur in the feet, it is not advisable to attempt the treatment of chronic laminitis, unless it is in valuable breeding animals.

QUESTIONS

1. Give a general description of the foot.

2. State the nature and causes of side-bones.

3. What are the causes of navicular disease? Give symptoms and treatment

4. What are corns? Give the treatment.

5. Give the nature and treatment of quarter- and toe-cracks.

6. Give the symptoms and causes of laminitis.

7. Give lines of treatment to be followed in the different forms of laminitis.

8. How may laminitis be prevented?

CHAPTER XVIII
DISEASES OF THE HIND LIMB

FRACTURE OF THE ILEUM, "Hipped."—Fracture of the angle and neck of the ileum may be classed among the common fractures in horses and cattle. Fractures involving other parts of the pelvic bones are less common. Such fractures are due to accidental causes, as striking the point of the haunch on the door frame when hurrying through a narrow doorway and falling on frozen ground.

Fractures of the *external angle* of the ileum or point of haunch are usually followed by displacement of the fractured portion. The same is true of fractures of the *neck of the ileum*. The result is a deformity of the quarter.

In making an *examination* of these parts the examiner should see that the horse is standing squarely on its feet, and then compare the conformation of the two quarters. Fractures of either the external angle or the neck of the ileum cause the quarter to appear narrow and low. A close examination may enable the examiner to differentiate between the two fractures. Fractures of the neck of the ileum can be recognized by manipulating the part through the walls of the rectum or vagina.

The degree of lameness may vary. In some cases there may be no lameness when the animal walks, but a slight degree of lameness may be noticed when it trots. For several weeks after the injury the horse may be unable to use the limb, but it may eventually make nearly a complete recovery.

Atrophy of the muscles of the hip or quarter should not be mistaken for fractures of the ileum. This condition involves the heavy gluteal muscles and may occur as a complication of azoturia, or a lameness of the hind limb that is usually due to a spavin.

It is very seldom necessary to give fractures of the ileum any special care. If the animal is very lame, it should be given a narrow stall, and placed in a sling until it can support its weight on the limb. The same treatment is indicated in cattle. It is not advisable to breed a mare that has had the ileum fractured. The bony enlargement that results from the union of the broken ends of the bone may interfere with the passage of the foetus through the pelvic cavity and cause difficult parturition.

LUXATION OF THE PATELLA, "Stifle Out."—This is a common accident in horses and mules. Young, immature animals are more prone to displacement of the patella than when mature. The displacement is usually upward or outward. Outward displacement is comparatively rare.

The causes of "stifle out" may be described as follows: The patella or knee-cap rests on a pulley-like articular surface belonging to the inferior extremity of the thigh-bone. The external lip of this articular surface is smaller than the internal lip. The patella is held in place from above by the heavy muscles of the anterior region of the thigh, and from below, by straight ligaments that attach it to the leg-bone. If the retaining structures mentioned become relaxed, the patella may, when the limb is extended, become so displaced as to rest on the superior portion of the external lip. Laxness of the muscles and ligaments in young animals is a predisposing factor. Hard work that tires the muscles and causes them to become relaxed, strains, unusual movements, as kicking in the stable and slipping, may cause this accident. Congenital displacement results from imperfect development of the external lip of the trochlea. Such a deformity subjects the animal to frequent luxations.

The symptoms may vary. The displacement may be first noticed when the horse is backed out of the stall or turned quickly. A slight "hitch" in the movement of the limb is noted, that is followed by more noticeable flexion of the hock than normal. In case the luxation is more permanent, the horse stands quietly with the affected leg held stiffly and extended backward. When made to move forward, it hops on the well leg and carries the affected one, or drags it on the toe. If both limbs are affected, the animal is unable to move. The inability to move the limb is due to the patella resting on the external lip of the pulley surface, and a locking of the stifle- and hock-joint.

This accident is annoying, and in case the horse is subject to it should be considered an unsoundness.

The following *treatment* may be recommended: The luxation may be reduced in the large majority of cases by backing or turning the animal. If this does not reduce the displacement, a collar should be placed on the animal, and a hobble strap fastened to the pastern of the involved limb. One end of a long rope is tied to the collar, passed backward between the front limbs, through a ring in the hobble and back over the outside of the shoulder and under the collar. While an attendant pulls the limb a little forward with the rope, the operator takes hold of the foot and attempts to flex the limb, at the same time pushing inward on the patella. After reducing the luxation it is advisable to tie the rope to the collar, so that the limb is carried forward. This prevents the animal from throwing weight on the foot. It may be advisable to tie the animal so that it can not lie down, if the foot is to be left hobbled for a few days. A fly blister should be applied to the front and outside of the stifle and the application repeated in two or three weeks.

STRING-HALT.—This term is applied to a peculiar involuntary movement of one or both hind limbs that is characterized by a sudden, purposeless flexion of the hock-joint . Horses that are slightly affected may show this movement of the hind limbs when first exercised. Other horses may be "string-halted" when backed, turned, walked, or trotted, and fail to drive out of it. The cause of true "string-halt" is not known.

The treatment recommended is surgical. This consists in cutting the tendon of the peroneus muscle. The seat of the operation is a little below the hock and on the external face of the cannon.

SPAVIN.—A spavin is a chronic inflammation of the articular faces of the hock bones, ligaments and synovial membranes. The inflammation may result in the formation of a bony enlargement on the inner surface of the region, and a union between the small bones forming the lower portion of the hock, and the upper extremities of cannon and lower hock bones (Figs. 46 and 47).

The *predisposing causes* are of the greatest importance. A spavin is one of the unsoundnesses of horses that may be transmitted to the offspring. Young colts that have heavy bodies and are fed a fattening ration are predisposed to it. Crooked hind limbs, small hocks and quarters that are heavily muscled are predisposing factors. The *external causes* are strains caused by slipping, turning quickly, rearing, pulling heavy loads and kicks. Horses three or four years of age if given work that favors hock strain, such as excavating cellars, may develop a spavin.

The symptoms or lameness are more characteristic than in most diseases of the limb. At the very beginning of the inflammation, and sometimes for several months afterward, the lameness is intermittent and disappears with exercise. After a time it is permanent. It is characterized by a stiffness of the hock. The extension of the hock is incomplete, the step is short and quick, the animal "goes on its toe" and the wall or shoe at the toe shows considerable wear. Because of the stiffness in the hock the animal raises the quarter when the limb is carried forward. Turning toward the well side may increase the lameness. The *spavin test* may be of value in diagnosing lameness. This consists in picking up the foot and holding the hock in a flexed position for a few minutes. The foot is then dropped to the ground and the animal moved off at a brisk trot. If the lameness is more marked, it indicates that the seat is in the region of the hock. This test is of greatest value in young animals. The bony enlargement can usually be seen best if the examiner stands in front and to one side of the animal. The hock should be observed from directly behind as well. The hocks of both limbs should be compared, and the general conformations of the other joints as well. This may prevent the examiner from mistaking rough hocks for spavin enlargements or "a pair" of spavins for rough hocks. A bony enlargement does not always accompany the lameness, and a spavin may be present without the horse going noticeably lame.

The prognosis is always uncertain and should be guided somewhat by the conformation of the limb, character of the work required of the animal, position of the bony enlargement and the degree of lameness. The size of the enlargement is changed very little by the treatment. Veterinarians report recoveries in from fifty to sixty per cent of the cases treated.

The object of the treatment is to destroy the inflammation and bring about a union between the bones. The treatment recommended is counterirritation and rest. The most satisfactory method of counterirritation is firing followed by blistering. Following this treatment, the horse should be placed in a stall and given no exercise for a period of five or six weeks. It is sometimes advisable to repeat the counterirritation if the results of the first firing are unsatisfactory.

BOG SPAVIN.—Bog spavin is an extensive distention of the capular ligament of the hock-joint by synovia . It is generally due to chronic inflammation of the synovial membrane. This blemish or unsoundness is most common in young horses. Thorough pin involves the sheath of the large tendon only. (Compare Figs. 48 and 49.)

Certain conformations of the hock favor the development of bog spavin. This is especially true of upright and "fleshy" hocks. Hard work may cause the hocks to "fill" when followed by a brief period of rest. The common cause is a sprain due to slipping and pulling heavy loads.

The *following symptoms* may be noted: Lameness is not a common symptom of bog spavin. If there is inflammation present or the articulation is injured, lameness occurs. The soft swelling that characterizes the bog spavin is most prominent toward the inside and front of the region. In the upper portion or hollow of the hock, and on the inside and outside, there may be a second enlargement. Smaller enlargements may be present in other regions. All of the swellings feel soft, and pressure on any one of them moves the fluid present in the others.

The treatment is directed at the removal of the lameness. Acute inflammation resulting from spavin may be relieved by cold applications and rest. Chronic lameness should be given the same treatment as recommended for bone spavin. The enlargement can be successfully removed in growing colts by the repeated application of mild blisters. It may be necessary to continue the treatment for several months. The removal of the enlargement in adult horses by an operation is recommended. The *greatest caution* is required in performing this operation.

CAPPED HOCK.—All swellings on the point of the hock are termed "capped hock." The swellings may be due to an injury to the skin and the subcutaneous tissue, or more important structures may be involved, as the subcutaneous bursa, the tendon, or the synovial bursa or sack.

Capped hock is *caused* by the animal kicking in the stall or in harness, shipping in freight cars and lack of bedding in the stall. Unless the deeper structures are bruised and inflamed the animal shows no lameness.

The character of the enlargement varies. When the injury is superficial, the swelling feels firm, or pits on pressure. Later it may become more firm and feel like a loose, thickened, fibrous cap for the hock. Soft, fluctuating swellings are due to an inflammation of the bursa. Recent injuries feel hot.

The preventive treatment consists in hobbling the hind limbs of a horse that kicks in the stable. This is usually necessary only at night. It may be advisable to pad certain parts of the stall. Horses that are transported in cars should be protected against injuries during transit by the use of proper care and such arrangement of the animals in the car as may expose them to the least injury. Recent injuries should be treated by the application of cold and rest.

After the inflammation has subsided tincture of iodine or blisters may be applied. The treatment of bursal enlargements is surgical. This consists in opening the bursa, destroying the lining membrane of the cavity and treating the part daily until healed. The operation must be performed carefully, as there is danger of infection with irritating organisms. The animal should be given complete rest until the part is healed. Tincture of iodine may be applied to the enlargement that may remain after healing has occurred. This should be continued daily until the skin becomes noticeably irritated. The treatment may be repeated, if necessary, after an interval of two weeks.

CURB.—This term is applied to all swellings on the posterior border of the hock . Thickenings or enlargements in this region may involve a variety of structures. Thickening of the skin, tendons and sheath may occur. The large ligament that extends from the posterior border of the bone that forms the summit of the hock to the external splint bone, and acts as a stay for the point of the hock, is the structure usually involved in curb.

Faulty conformation is a *predisposing cause*. A narrow base weakens the hock at this point, and the extreme length of the bone that forms its summit gives the powerful muscles attached to it greater leverage than in a well-conformed hock. This results in strain to the ligament at the posterior portion of the region.

The exciting causes are strains resulting from jumping, slipping, rearing, heavy pulling and bruising of the part.

In examining the hock for curb it is necessary to stand to the side and note the profile of the posterior border. Excessive development of the head of the external splint bone should not be mistaken for curb. As viewed from the side, the posterior border of the hock should appear straight.

The curb appears as a swelling on this straight line. It varies in size. A recent curb is usually hot and firm, or may feel soft if enlargement is formed by fluid, hard if formed by bone. Lameness seldom occurs, but if present, resembles spavin lameness.

The preventive treatment consists in selecting for breeding, animals that have strong, straight hocks, and using the necessary care in handling and working horses. It is not uncommon for young horses at the time they are broken to harness to develop a curb. This may be prevented to a large degree by careful handling. At the beginning of the inflammation the application of cold and hand rubbing is indicated. After the inflammation has subsided tincture of iodine or blisters should be applied. Rest is a necessary part of the treatment early in the inflammation. If the lameness does not respond to the above treatment, it should be treated the same as for bone spavin.

QUESTIONS

1. Describe the different fractures of the ileum and give treatment.

2. Describe string-halt lameness and give treatment.

3. What is bone spavin? Describe spavin lameness.

4. Give the causes and treatment of bog spavin.

5. Give the causes and treatment of capped hock.

6. Give the causes and treatment of curb.

PART III
THE TEETH

CHAPTER XIX
DETERMINING THE AGE OF ANIMALS

GENERAL DISCUSSION. — The teeth are the passive organs of digestion. They are hard organs, implanted in the superior and inferior jaws in the form of a long and narrow arch that is open posteriorly. The free portions of the teeth project into the mouth, and present sharp or roughened table surfaces for the crushing and tearing of food. In solipeds and ruminants the arch is interrupted on each side by the inter-dental space or bars . The teeth that form the middle and anterior portion of the arch are termed incisors . Posterior to the incisors are the canines or tusks, and forming the arms of the arch are the molar teeth. Animals have two sets of teeth, temporary and permanent. The following table gives the number of the different kinds of temporary and permanent teeth.

Temporary Teeth Permanent Teeth
Incisors Canines Molars Incisors Canines Molars

Solipeds 12 12 12 4 24
Ox 8 12 8 0 24
Sheep 8 12 8 0 24
Hog 12 12 12 4 24

The tusks or canine teeth are not always present in the female. Ruminants do not have upper incisor teeth. The temporary teeth are erupted either before or within a few days to a few months after birth. The eruption of the permanent teeth and the replacement of the temporary teeth occur at different periods up to the age of four and one-half years . It is well to keep the following table of dentition in mind when examining the mouths of animals for the purpose of determining their age.[1]

Horses Cattle Hogs
Teeth Temporary Permanent Temporary Permanent
Temporary Permanent

Incisors: yrs. mos. yrs. mos. mos.
Centrals At birth 2 6 At birth 1 8 At birth, 12
or 3-4
weeks
First 4-6 wks. 3 6 At birth 2 9 8-12 wks. 18
 laterals
Second 5-12 days 3 6
 laterals
Corners 6-9 mos. 4 6 12-18 days 4 6 At birth 9
Molars:
First At birth 2 6 At birth 2 6 7 weeks 5
Second At birth 2 6 At birth 1 6 8-28 days 14
Third At birth 3 6 At birth 3 8-28 days 13
Fourth 10-12 1 6 13
Fifth 2 2 5
Sixth 4-5 2 6 9
Seventh 18
Canines or 4-5 9
 tusks

IN DETERMINING THE AGE of the different domestic animals by the development and appearance of the teeth, most of the attention is given to the lower incisor teeth. Up to the fifth year, the age of the horse or ox can be easily determined by the eruption and replacement of the incisors.

At *one year* of age the colt has a fully developed set of temporary incisors. The ruminant's incisors at this age all show wear.

The two-year-old colt shows a well-worn set of incisor teeth, and the ruminant at this age has replaced the nippers or centrals.

The third, fourth and *fifth years* are indicated by the replacement of the temporary nippers, dividers and corners in the horse, and the first and second dividers and corner teeth in ruminants.

In the horse the permanent nippers are full grown and in wear at *three years* of age; the permanent dividers are full grown and in wear at *four years* of age; and the permanent corners are full grown and in wear at *five years* of age. The table surfaces of the incisor teeth of a five-year-old horse show different degrees of wear. At this period in the animal's age, the nippers have been in wear two years, the dividers one year, and the corners are beginning to show wear. In ruminants, all of the chisel-shaped table surfaces of the incisors show considerable wear when the animal is five years old.

After the animal has a full set of permanent teeth, we judge the age by the degree of wear or the appearance of the table surfaces of the incisors, their shape, the angle with which they meet and the general appearance of the head.

There are several different factors that may cause the wear on the teeth, and the appearance of their table surfaces to vary in the different individuals. The two factors that are of the most importance are the quality of the teeth and the character of feed. Soft teeth wear more quickly than hard teeth, and the teeth of horses that feed over closely cropped and sandy pastures wear rapidly because of the dirt and grit present on the short grass. This variation in the wear is of little importance to the person who must judge the age of a horse that he expects to purchase by the condition of the teeth. In reality, a horse is just as old as the wear on the teeth and his general appearance indicate. In order to stand severe work the animal must be able to masticate the feed, and prepare it for digestion in the stomach and intestines. The degree of wear on the molar teeth may be indicated by the wear on the incisors. The general condition of the horse and his ability to stand hard work depend very largely on the condition of the table surfaces of the molars.

It is very difficult to judge the age of horses that have deformed mouths or that are in the habit of crib-biting, because of the irregularity in the wear of the incisors.

When examining the teeth for the purpose of determining the horse's age, the shape of the incisors, the angle with which they meet and the appearance of their table surfaces should be observed. The teeth of young horses show more or less yellowish cement. At about seven years of age the anterior faces of the teeth are usually white, later a yellowish color. The teeth of middle-aged horses may be long, and in aged animals, narrow and short. The incisors meet at a more acute angle in old than young horses.

The free portion of the incisor tooth is flattened from before to behind. At the level of the gums its two diameters are about the same, but the portion of the tooth imbedded in the jaw bone is flattened from side to side. As the tooth becomes worn off, the length of the free portion is maintained by a pushing out of the tooth, and a corresponding shortening of the portion that is fixed or imbedded in the jaw.

The table surface of the unworn incisor tooth is covered with enamel, and in the middle portion the enamel forms a deep cup. After the tooth has become worn the margin of the table portion is then limited by a ring of enamel. This is termed the encircling enamel ring. The central portion of the table shows a second ring, the central enamel ring, that limits the cup margin .

As the table surface represents a cross section of the tooth, its appearance and shape will then depend on the portion of the tooth that it represents. From year to year, there is a gradual shortening in the lateral diameter, and an apparent increase in the diameter from before to behind. These changes in shape are from a long, narrow table surface to an oval, from oval to circular and from circular to triangular . As the original free portion of the tooth wears off, the cup becomes shallow and smaller until the remnant is represented by a mere dot of enamel that finally disappears from the posterior portion of the table. After the cup has moved from the central portion of the crown and occupies a more posterior position, the dental star, which represents a cross section of the pulp cavity, puts in its appearance. It first takes the form of a brown or dark streak, and later a circular dark spot which gradually increases in size with the wear on the tooth and the age of the animal.

The following changes in the shape and appearance of the incisor teeth of the average horse occur in the different years. Unless otherwise mentioned, the statements made regarding the appearance and wear on the table surfaces apply to the lower incisor teeth.

SIX YEARS.—The table surfaces form the most accurate guide. The cups of the nippers tend to an oval form. The corner teeth have been in wear one year at this time. The cup is deep and the posterior margin may show little wear. It is not uncommon to meet with corners that possess irregularly developed tables, and have cups with posterior margins that are thin and do not come into wear until later. For this reason, it is not best to depend on the appearance of the corner teeth alone.

SEVEN YEARS.—The teeth are usually whiter than the previous year. The profile of the upper corner teeth shows a notch in the posterior portion of the table surface. This is due to the superior corners overhanging the inferior corner teeth posteriorly, resulting in this portion not wearing away. This notch is sometimes slightly in evidence the previous year. The cups in the corners are smaller and the worn surface larger than at six. The nippers show oval table surfaces and the dividers are beginning to take on this shape. The shifting of the cups toward the posterior portion of the tables of the nippers and dividers is noticeable.

EIGHT YEARS.—As viewed from the side, the profile of the teeth shows a very noticeable increase in the obliquity with which they meet. The posterior borders of the corners show considerable wear. The notch in the superior corners is still present, but as the teeth come more nearly in apposition it may begin to disappear. All of the inferior tables are level. The nippers and dividers are oval in shape, and the cups have become decidedly narrow. The nippers show a well-defined dark streak just in front of the cups. This is the beginning of the dental star.

NINE YEARS.—The appearance of the table surface is more characteristic at this time than the previous year. The cups are less prominent and the plainness or smoothness of the inferior table is more noticeable. The nippers are round, the cups triangular and the dark streak narrower and more distinct than the previous year. The dividers are becoming round and the corner teeth are oval.

TEN YEARS.—The teeth are more oblique than in the eight-year-old and nine-year-old mouth. The table surfaces of the inferior nippers are decidedly rounded, the cups are small, triangular and situated well toward the posterior borders. The dark brown streak or dental star is situated in the central portion of the nippers and dividers. The tables of the dividers are round.

ELEVEN YEARS.—The tables of the corner teeth are rounded. The dark streak or dental star is present in all of the teeth, and the remnants of the cups appear as small rings or spots of enamel near to the posterior borders of the tables. The notch in the superior corners may reappear at this time.

TWELVE YEARS.—The profile of the teeth when viewed from the side is quite oblique. The table surfaces of all the incisors are round. But a trace of the cup remains in the inferior incisors. The head of the animal is beginning to show age. The inferior border of the jaw bone appears narrower, or sharper than in the young horse.

THIRTEEN YEARS.—All of the specks of enamel or the remnants of the cups are gone from the lower incisors. A larger notch may be present in the upper corner teeth than at twelve. The tables of the inferior nippers are becoming triangular and show a small, dark spot or dental star.

FOURTEEN YEARS.—The tables of the inferior nippers are triangular, and the dental star appears as a dark round spot in both the nippers and dividers.

FIFTEEN YEARS.—The angle with which the teeth meet is greater than at twelve, the teeth are smaller and dental stars are represented by dark round spots in all of the inferior incisors. The tables of the nippers and dividers are triangular.

SEVENTEEN YEARS.—All of the tables of the lower incisor teeth are triangular. The teeth are narrower and smaller than at fifteen. The profile of the incisors, viewed from the side, is quite angular. The dental stars are prominent.

NINETEEN YEARS.—All of the signs of the seventeen-year-old mouth are more prominent. The cups have usually disappeared from the upper incisors.

[Footnote 1: This table is from dentition tables given in "Age of the Domestic Animals," by Huidekoper.]

QUESTIONS

1. Name the different kinds of teeth; state the arrangement and number.

2. How is the age of an animal determined?

3. Give the time of replacement of the temporary incisor teeth.

4. How is the age of the animal determined between the fifth and ninth years?

5. What changes in the appearance of the table surfaces occur between ten and fifteen years of age?

CHAPTER XX
IRREGULARITIES OF THE TEETH

Parrot-mouth, Lantern-jaw and Scissor-mouth.—The common deformities of the jaw and teeth are the overshot or parrot-mouth, the undershot or lantern-jaw, and the scissor-mouth. These different deformities result in unequal wear on the table surfaces of the incisors and molars. In both the overshot and undershot jaws, the incisor teeth become abnormally long. In the *parrot-mouth,* the wear occurs on the posterior face of the superior and the anterior face of the inferior incisors, the teeth becoming worn to rather a sharp edge, depending on the degree of the deformity. In the *lantern-jaw,* the wear occurs on the posterior face of the lower and the anterior face of the superior row of incisors, the teeth taking on somewhat the same shape as the parrot-mouth. The greater the deformity and the older the horse becomes, the more difficult it is for the animal to feed or graze on pasture.

In all horses, the two rows of molar teeth are wider apart in the superior than in the inferior jaw. This results in the external border of the tables of the superior row of molars becoming longer, or projecting further downward than the internal border. The wear on the table surfaces of the inferior row of molars is just the opposite of the superior row. In the *scissor-mouth* the wear takes place largely on the internal face of the superior and the external face of the inferior row of molars. The teeth become worn to more or less of a blunt cutting edge, and after a time the molars come together somewhat like the jaws of a pair of scissors. A horse with a badly deformed scissor-mouth is unable to grind the feed, and unless given special care, suffers severely from innutrition.

The treatment of deformed mouths consists in removing the irregular or unworn portion of the teeth by means of the tooth float and cutters. This attention should be given early before the free portion of the tooth has become excessively long and irregular. This should be followed by dressing the teeth every six or twelve months.

SHARP LATERAL BORDERS ON THE MOLAR TEETH.—This is a very common condition in horses. The external border of the superior and the internal border of the inferior row of molars wear away slowly, and sometimes become quite sharp. This is objectionable because the sharp points lacerate the mucous membrane of the cheek and tongue, and the mastication of the feed is seriously interfered with.

This condition is *caused* by an excessive difference in the width of the jaws, unusually prominent ridges of enamel on the external face of the superior molars, and any conditions that may limit the movements of the jaw.

The following *symptoms* may be noted. The animal has difficulty in masticating the feed because of injury to the cheeks or tongue by the sharp points of enamel. This condition may be indicated by holding the head to one side. Salivation is usually present. Acute indigestion and innutrition may occur.

By examining the teeth, their condition can be determined. The sharp borders may be removed by dressing or floating the teeth. It is advisable in the majority of horses to float the teeth at least once in every twelve months.

IRREGULARITIES IN THE TABLE SURFACE OF THE MOLAR TEETH.—Horses eight years of age or older frequently have irregular molars . This is due very largely to the difference in the quality of the teeth. The harder molars do not wear off as rapidly as the softer ones. This results in the table surfaces of the rows of molars becoming wavy or step-like in outline. Sometimes the first or sixth molar overhangs or projects beyond the corresponding tooth of the opposite jaw. When this occurs, the over-hanging portion may become long and sharp. A molar tooth becomes excessively long if the opposite one is decayed or removed.

The symptoms are very much the same as when the borders of the molars are sharp.

The treatment consists in levelling the tables as frequently as necessary by cutting off the longer projections, and removing the sharp edges with a tooth float.

SMOOTH MOUTH.—In old age the tables of the molar teeth may become so smooth that the horse cannot grind or masticate the feed. When all of the molars are in this condition, a rubbing sound may be noted when the animal is masticating hay. After attempting to chew the hay, it may be dropped from the mouth. Innutrition always occurs.

The treatment consists in feeding chops and soft feeds.

DENTAL DISEASES.—Inflammation of the alveolar periosteum is a common dental disease in domestic animals. This is an inflammation of the alveolar dental membrane that fixes the tooth in the tooth cavity.

Injuries to the gums and cracks or fissures in the tooth are the *common causes*. Caries or tooth decay is not uncommon. The predisposing factor is a poor quality of enamel and dentine. The process of decay is assisted by microorganisms.

The *early symptoms* may escape notice. Slobbering, masticating on one side, holding the head to one side, retained masses of food in the mouth and a disagreeable odor frequently occur. Caries may be indicated at first by a dark spot on the table of the tooth, later by a cavity. In horses, inflammation of the alveolar membrane results in a bony enlargement on the side of the face if the superior molar is involved. A swelling of the jaw and fistula may result if a lower molar is involved .

The treatment consists in the prompt removal of the tooth. This is more difficult in young animals than it is in the middle-aged or old. Unless the tooth is already loosened it may be necessary to remove it by trephining.

QUESTIONS

1. Describe the appearance of the teeth in an overshot or undershot jaw.

2. Describe the appearance of sharp molar teeth; a scissor-mouth.

3. What are the causes of decayed teeth?

PART IV
SURGICAL DISEASES

CHAPTER XXI
INFLAMMATION AND WOUNDS

Inflammation is a pathological condition of a tissue, characterized by altered function, disturbance of circulation, and destructive and constructive changes in the irritated part. Heat, redness, swelling, pain and disturbed function are the symptoms which characterize inflammation.

The changes in the circulation occurring in inflammation are as follows: (1) An increase in the rate of the blood-flow through the blood-vessels of the part and their dilation; (2) diminished velocity followed by the blood-flow becoming entirely suspended; (3) following the retardation or suspension of the blood stream, white blood-corpuscles accumulate along the walls of the small veins and capillaries; (4) white and red blood-corpuscles migrate from the vessels into the neighboring tissue, and blood-serum transudes through the walls of the vessels, forming the inflammatory swellings. The red blood-cells do not escape from the blood-vessels in any numbers unless the walls of the blood-vessels become injured or badly diseased.

The causes of inflammation may be grouped under the following heads: mechanical, chemical, thermic and infectious. The *mechanical* or *traumatic causes* commonly produce inflammation in domestic animals. These are kicks, strains of tendons, ligaments or muscles and wounds. Inflammation originating from injuries very frequently changes to an infectious form, through the infection of the part by bacteria. Bruised tissue may become infected with pus-producing organisms, and an abscess or local swelling form. All accidental wounds in domestic animals become more or less infected by irritating microorganisms.

The following *symptoms* occur in local inflammation. Increased heat in the part is an important symptom. It is due to the increased blood-flow to the part. Because of the pigmented, hairy skin of domestic animals, redness is of little value in locating superficial inflammation. Swelling is a valuable local symptom. It is produced by the inflammatory exudates. Pain results from the pressure on the sensory nerves by the inflammatory swelling. For example, the laminae of the foot are imprisoned between the horny wall and the pedal bone. This structure is well supplied with sensory nerves, and when it becomes inflamed and swollen, the tissues are subject to severe pressure and the pain is severe. Inflammation of a tendon results in lameness; of the udder, in suspension of milk secretions; and of the stomach by interference with digestion of the feed. Such symptoms may be grouped under the head of disturbed functions.

The character of an inflammation is largely modified by the nature of the tissue in which it occurs. A serous inflammation is characterized by serous, watery exudates. This form occurs in the serous membranes, mucous membranes and skin. Blisters on the skin and inflammation of bursae (capped hock and shoe boil) are examples of this type. Sero-fibrinous inflammations, such as occur in pleurisy and peritonitis, are common. Chronic inflammation commonly results in new formations of tissue, and it is named according to the character of the new tissue formed, as ossifying, adhesive, and fibrous inflammation. Pus-forming bacteria produce suppurative inflammation. Such diseases as tuberculosis, glanders and hog-cholera are specific inflammations. Specific infectious diseases may be classed as generalized inflammation, as they usually involve the entire body.

Inflammation terminates in resolution when the serum is reabsorbed by the blood-vessels and lymphatics, the living blood-cells find their way back into the circulation and the dead cells disintegrate and are taken up by the vessels. The time required for the tissues to return to the normal varies from a few hours to several weeks. An acute inflammation may end in the chronic form. This may then terminate in new formations, such as adhesions, fibrous thickenings and bony enlargements. Severe inflammation, especially if localized and superficial, may result in death of the part or gangrene.

The following *treatment* is recommended: The cause of the irritation to the tissue must be removed. It is very essential that the part be rested. The necessary rest may be obtained in different ways. Inflamed tendons, ligaments, and muscles may be rested by placing the animal in a sling, standing it in a stall, or fixing the part with bandages. Rest of the stomach or intestinal tract may be obtained by feeding a light diet, or withholding all feed. Comfortable quarters, special care and dieting the animal are important factors in the treatment of inflammation.

The agents used in the treatment of superficial and localized inflammation are *heat, cold, massage* and *counterirritation. Heat* is indicated in all inflammations, excepting when of bacterial origin. It stimulates the circulation and reabsorption of the inflammatory exudates, and by relaxing the tissues helps greatly in relieving pain. *Cold* is more effective in the highly acute and septic (suppurative) inflammation. Its action consists principally in the contraction of the dilated blood-vessels. Continuous irrigation of the part with cold water is the most satisfactory method of applying cold. *Massage* is a very important method of treating superficial inflammation. Mild, stimulating liniments are usually used in connection with hand-rubbing or friction. Chronic inflammation is usually treated with *counterirritants*. Blistering and firing are the most important methods of treatment. Such counterirritation makes possible the absorption of the inflammatory exudates by changing the chronic inflammation to the acute form.

WOUNDS.—A wound, in the restricted sense that the term is commonly used, includes only such injuries that are accompanied by breaks or divisions of the skin and mucous membrane. It is usually an open, hemorrhagic injury.

If the tissues are severed by a sharp instrument and the edges of the wound are smooth, it is classed as an *incised* or *clean-cut wound*. This class is not commonly met with in domestic animals outside of operative wounds.

When the tissues are torn irregularly, the injury is classed as a *lacerated wound*. A barb-wire cut is the best example of this class.

A *contused wound* is an injury caused by a blunt object. Such injuries may be divided into superficial and deep. Superficial-contused wounds may be an abrasion to the skin or mucous surface. Deep-contused wounds may be followed by loss of tissue or sloughing, and may present irregular, swollen margins. Such injuries are commonly caused by kicks.

Punctured wounds are many times deeper than the width of the opening or break in the skin or mucous membrane. This class is produced by sharp objects, such as nails, splinters of wood, and forks.

Sometimes, wounds are given special names, as gun-shot, poisoned, and open joint, depending on the nature of the cause and region involved.

Bleeding or hemorrhage is the most constant symptom. The degree of hemorrhage depends on the kind, number and size of the blood-vessels severed. In arterial hemorrhage, the blood is bright red and spurts from the mouth of the cut vessel. In venous hemorrhage, the blood is darker and flows in a continuous stream. In abrasions and superficial wounds capillary hemorrhage occurs. Death may follow severe hemorrhage. Weak pulse, general weakness, vertigo, loss of consciousness and death may result if one-third of the total quantity of blood is lost. Unthriftiness and general debility may follow the loss of a less quantity of blood.

The following *symptoms* may be noted in the different kinds of wounds: The sensitiveness to the pain resulting from accidental or operative wounds varies in the different individuals and species, and in the kind of tissue injured. Injuries to the foot, periosteum, skin and mucous membrane are more painful than are injuries to cartilages and tendons. The appearance of the wound varies in the different regions and the different tissues.

If the tissues are badly torn or bruised, swelling and sloughing may occur. If the wound is transverse to the muscular fibres, it gaps more than when parallel to the muscle. When infected by irritating organisms, open and punctured wounds become badly swollen, discharge pus freely and heal slowly with excessive granulations. Wounds involving tendons, bursae and closed articulations become swollen and discharge synovia. Wounds involving muscles, tendons and bursae usually cause lameness, and when involving a special organ, interfere with, or destroy, its function. Extensive or serious wounds may be followed by loss of appetite. An abnormal body temperature and other symptoms characteristic of the different forms of blood poisoning may follow infection of the injured tissues by certain germs.

The rapidity with which wounds heal depends upon the kind of tissue injured and the amount to be replaced, the degree of motion in the part, the kind and degree of infection and irritation and the general condition of the animal. In general, skin and muscles heal rapidly, tendons slowly, cartilages unsatisfactorily and nerve tissue very slowly. Healing is greatly interfered with by movement of the part . The more nearly the part can be fixed or rested, the more quickly and satisfactorily does healing occur. Irritation by biting, nibbling, licking, bandaging, wrong methods of treatment and filth retard healing and may result in serious wound complications. An animal in poor physical condition, or one kept under unfavorable conditions for healing, cannot recover from the injury rapidly or satisfactorily.

WOUND HEALING.—The following forms of healing commonly occur in wounds:
First and second intention; under a scab, and by abnormal granulation.

Healing by first intention occurs when the wound is clean cut and there is very little destruction of tissue, and when there is no suppuration or pus formation. The blood and wound secretions cause the edges of the wound to adhere. After a few days or a week the union becomes firm. Very little scar tissue is necessary in this form of healing.

Healing by second intention is characterized by pus formation and granulation tissue. After the first day, the surface of the wound may be more or less covered by red, granular-like tissue. Later this granular appearance is modified by an accumulation of creamy pus and swelling of the part, and finally scab formation and contraction of the new scar tissue.

Abrasions and superficial wounds usually *heal under a scab*. The scab is formed by the blood and wound secretions. This protects the surface of the wound until finally the destroyed tissue is replaced by the granulations, and the skin surface is restored.

Abnormal granulation is not an uncommon form of healing in domestic animals. Mechanical and bacterial irritation causes the injured tissue to become swollen and inflamed. In such a wound, excessive and rapid granulation occurs, the new tissue piling up over the cut surfaces and appearing red and uneven. This is termed excessive granulation or "proud flesh." This tissue may refuse to "heal over," or the scar may be large, prominent and painful. Abnormal tissue (horny or tumor-like) may sometimes form.

WOUND TREATMENT.—Wounds in domestic animals are frequently allowed to heal without special care or treatment. This is unfortunate. The careful and intelligent treatment of wounds would greatly decrease the loss resulting from this class of injuries. The method of treatment varies in the different kinds of wounds.

The first step in the treatment is to *check the haemorrhage*. Heat, ligation, pressure and torsion are the different methods recommended. Bathing the wound with hot water (115\260-120\260 F.) is a satisfactory method of controlling haemorrhage from small blood-vessels. Ligation and torsion of the cut end of large blood-vessels should be practised. Pressure over the surface of the wound is the most convenient method of Controlling haemorrhage in most cases. Whenever possible, the part should be bandaged heavily with clean cheese cloth or muslin. Before applying the bandage, it is advisable to cover the wound with a piece of sterile absorbent cotton that is well dusted with boric acid. Hemorrhage from wounds that cannot be bandaged may be temporarily stopped by pressure with the hand, or, better, by packing the wound with absorbent cotton and holding this in place with sutures. This should be left in place for a period of twelve or thirty-six hours, depending on the extent of the haemorrhage and character of the wound.

The next step is the *preparation of the wound for healing.* The injured tissues should be carefully examined for foreign bodies such as hair, dirt, gravel, slivers of wood and nails. The hair along the margins of the wound should be trimmed, and all tissue that is so torn and detached as to interfere with healing cut away. Drainage for the wound secretions and pus should be provided. The advisability of suturing the wound depends on its character and location. A contused-lacerated wound should not be closed with sutures unless it is clean and shows no evidence of sloughing. A badly infected wound should be left open unless satisfactory drainage for the pus and wound secretions can be provided. Wounds across the muscle and in parts that are quite movable should not be sutured.

The after-treatment consists in keeping the animal quiet, if the wound is in a part that is quite movable, and preventing it from biting, licking or nibbling the injury. Wounds in the region of the foot become irritated with dirt and by rubbing against weeds and grass. This makes it advisable to keep the animal in a clean stall until healing is well advanced. Local treatment consists in keeping the wound clean by washing the part daily, or twice daily, with a one per cent water solution of a cresol disinfectant. Liquor cresolis compositus may be used. It is sometimes advisable to protect the granulating surface against irritation by dusting it over with a non-irritating antiseptic powder, or applying a mixture of carbolic acid one part and glycerine twelve parts. After the wound shows healthy granulations longer intervals should lapse between treatments.

In poorly cared for, and badly infected wounds, the part may become badly swollen, the granulations pile up and the wound refuse to "heal over." It may be advisable in such cases to cut away the excessive granulations and stop the haemorrhage by cauterization with a red-hot iron, or by compression. Unhealthy granulations may be kept down by applying caustic occasionally.

ABSCESS.—This is an accumulation of pus in the tissues. It may be due to a severe bruise or contusion that is followed by the infection of the part with some of the pus-producing bacteria. Abscesses occur in certain infectious diseases. In strangles, the disease-producing organism may be carried to different regions of the body by the circulatory vessels. This may result in a number of abscesses forming in the different body tissues.

The following *forms of abscess* are recognized: hot and cold, superficial and deep, simple and multiple. The hot is the acute, and the cold the chronic abscess. The terms superficial and deep allude to the relative position of the abscess, and simple and multiple to the number present.

An abscess may first appear as a hot, painful swelling. If superficial, the skin feels tense and the contents fluctuate when pressed on. Later the fever subsides and no pain may occur when the abscess is pressed upon. Deep abscess may not fluctuate.

The treatment consists in converting the abscess into an open wound whenever possible. The incision should extend to the lowest part of the wall, so as to insure complete drainage. A cold abscess in the shoulder region may become lined by a layer of tissue that retards healing. In order to hasten the healing process, it may be necessary to remove this. Until granulation is well advanced, the abscess cavity should be irrigated daily with a one per cent water solution of liquor cresolis compositus, or a one to two thousand water solution of corrosive sublimate. The surface of the skin in the region of the abscess should be kept clean.

FISTULOUS WITHERS AND POLL EVIL.—These terms are applied to swellings, blood tumors, abscesses and pus fistulae that may be present in the region of the poll and withers . Pus fistula is the characteristic lesion present, and it is the result of a suppurative inflammation of the tissues in the region. The abscess cavity or cavities are usually deep, and may involve the ligaments and vertebrae.

Bruises or contusions are the most *common causes*. The prominence of these regions predisposes them to injury in the stable, or when rolling on rough or stony ground. Bites and bruises to the withers resulting from other horses taking hold of the region with the teeth, or striking the part against a hard surface, are frequent causes.

The treatment is both preventive and surgical. All possible causes should be investigated. This is of special importance on premises where several horses develop fistulous withers and poll evil. If the cause then becomes known, it should be removed.

The surgical treatment consists in opening up the different abscess cavities, providing complete drainage for the pus and destroying the tissue that lines the walls of the cavities. Horses that are prone to rub the region should be prevented from doing this, as such irritation retards healing. Autogenous bacterins should be used in addition to the surgical treatment. A pus fistula should heal from the bottom, and if the opening becomes closed, drainage should be re-established. The daily treatment is the same as recommended for abscesses. Excessive cutting and destruction of the tissues with caustic preparations result in scarring and deformity of the part. Such radical lines of treatment should be discouraged. We should not delay the surgical treatment of abscesses in the regions of the poll and withers.

QUESTIONS

1. Name and describe the different forms of inflammation.

2. Give the causes and treatment of inflammation.

3. Name and describe the different methods by which wounds heal.

4. Describe the treatment of wounds.

5. What are the causes of an abscess? Give the treatment.

6. What are the causes of fistula and poll evil? Give the treatment.

CHAPTER XXII
FRACTURES AND HARNESS INJURIES

FRACTURES.—Broken bones or fractures are not uncommon in domestic animals. In the horse, the bones of the leg, forearm, foot, and spine are the most commonly broken. In the dog the largest percentage of fractures occurs in the superior regions of the limbs.

Fractures may be classified as *simple* and *compound, complete* and *incomplete, comminuted* or *splinter*. In the simple fracture the skin over the region escapes injury, but in the compound fracture the skin is broken and the ends of the broken bone may protrude through it. The terms complete and incomplete are used in describing fractures in which the ends of the bones are not attached to each other, or partially so. In the comminuted fracture the bone is broken into a number of pieces. There are a number of other terms that may be used in designating the different kinds of fractures, such as double, when both bones in the region are broken, and oblique, transverse and longitudinal, depending on the direction of the break.

The causes of fractures may be divided into external or mechanical, and internal. Fractures may result from kicks, blows, muscular strain and contusions. Abnormal fragility due to disease, extreme youth and old age are the internal predisposing factors.

The symptoms are crepitation, abnormal movement and deformity of the part. Abnormal movement of the part and inability to support weight occur in fractures of the bones of the limbs. Crepitation or a grinding, rubbing sound due to the movement of the ends of the broken bones on one another occurs when the part is moved or manipulated with the hands. Pain, swelling and injury to the skin are other local symptoms. The new tissue or bone callus is formed by the bone-forming cells in the deeper layer of the periosteum and bone-marrow.

The prognosis is unfavorable. The larger percentage of fractures in domestic animals are incurable, or make an unsatisfactory recovery. This is due to careless treatment, the character of the fracture and the inability to fix the ends of the broken bone. Fractures in young and small animals usually heal quickly. Individuals that are healthy and vigorous usually make a speedy recovery. Fractures heal very slowly in the aged. Compound and comminuted fractures are impossible to treat in the larger percentage of cases.

The treatment consists in fixing the broken bone or bones in a normal position by means of bandages and splints. If this is not practised, the surrounding tissues become injured by the broken ends of the bone, and the fracture may become so complicated as to render treatment useless. Motion retards or prevents the repair of the break.

However, fractures of the ribs, pelvic bones and sometimes long bones that are well covered by heavy muscles heal naturally or in the absence of any means of retention.

Bandaging.—The attendant must use good judgment in devising means of fixing the broken bone, and in holding it in its natural position. Whenever possible, a plaster bandage should be used. This must not be made too heavy, and it is very necessary to adjust it properly, so that it will stay in place and not become too tight or too loose. When applied to the limb, the bandage should extend as far down as the hoof, and some distance above the break. This is necessary in order to keep it from slipping down and becoming too loose. A soft bandage should be applied first in order to equalize the pressure from the plaster cast and protect the skin. Wooden splints are not very satisfactory agents for the treatment of fractures. Thick leather that has been made soft by soaking in warm water and then shaping it to the part makes a more satisfactory splint. In all cases a soft bandage should be applied under the splint. The adjustment of the plaster bandage or splint should be noticed daily, and whenever necessary it should be removed and readjusted. Injuries to the skin must be carefully cleaned, disinfected and bandaged before applying the plaster bandage. If evidence of wound infection occurs later, the bandage must be removed and the wound treated. Large animals suffering with a fracture of any of the bones of the limb should be placed in slings. Incomplete fracture should receive the same treatment as simple fracture. If this is practised, the danger of its becoming complete is avoided.

HARNESS INJURIES.—This class of injuries is common in horses that are given steady, hard work, or that are not accustomed to work. Young horses, when first put to hard work, are especially prone to injuries from the collar. A large proportion of these injuries are due to an ill-fitting harness or saddle.

When the harness is not adjusted or fitted properly, there is severe pressure on certain parts. This is the common cause of shoulder abscesses , sore necks and sit-fasts. Rough, uneven surfaces on the faces of the collar and saddle are the common causes of galling. The character of the work is an important factor. Work that requires the animal to support weight on the top surface of the neck is productive of sore neck. Heavy work over rough, uneven ground frequently causes shoulder abscesses and strained muscles.

The simplest and most common harness injuries are galling, sore shoulders and sore neck. Harness galls first appear as flat, painful swellings. On raising the collar from the skin the inflamed area appears dry and the surrounding hair is wet with sweat. Later, the skin becomes hard and its outer layer, and sometimes the deeper layer as well, slough, or is rubbed off by friction of the harness. The surface then appears red and moist. Fluctuating swellings due to small collections of blood and lymph sometimes form. Sometimes, small areas on the face of the shoulder and that portion of the back pressed on by the saddle become swollen, indurated and hard and give the shoulder a rough appearance. Continuous irritation from the collar may cause an inflammatory thickening of the subcutaneous tissue in the shoulder region, and the skin appears loose and somewhat folded. This uneven surface is productive of chronic collar galls.

A sit-fast is characterized by a large swelling at the top of the neck, followed by a deep sloughing of the tissues. A slightly swollen, wrinkled condition of the skin over the top of the neck is sometimes present in horses that resist the attendant, when he attempts to handle the part or harness the animal. This form of sore neck is evidently very painful, although little evidence of inflammation is present.

Strain of shoulder muscles and shoulder abscesses have been discussed under their separate heads.

The treatment is very largely preventive. Too little attention is given to the proper fitting of the harness and saddle. A well-fitted collar that properly distributes the weight on the shoulder, and is neither too small or too large at the top of the neck, is the best preventive for shoulder and neck injuries. Old, ill-fitting, lumpy collars should not be used. Neither should the same collar be used for different horses. Farmers should avoid using sweat pads that are lumpy or soaked with sweat. If soft and dry, such pads are useful in preventing galling. The surfaces of the collar or saddle that come in contact with the skin should be kept smooth and clean. In the spring of the year, it is advisable to bathe the shoulders of work horses with cold water twice a day. Bathing the shoulders with the following preparation is a useful preventive measure: lead acetate four ounces, zinc sulfate three ounces and water one gallon. Smooth leather pads for the top of the collar and saddle are useful preventive and curative agents.

Galls are lest treated by rest. Ointments or "gall cures" are usually applied. The following dry dressing dusted over the red, moist, abraded surfaces is quite healing: tannic acid one ounce, boric acid four ounces, and calomel two ounces. This may be dusted over the part two or three times daily. Dry, abraded surfaces may be treated by applying a mixture of glycerine four ounces, tannic acid one-half ounce and carbolic acid one dram. In operating for the removal of fibrous enlargements, thickened skin and abscesses on the front of the shoulder, it is advisable to make the incision in the skin well to the side of the face of the shoulder in order to avoid scarring the surface that comes in contact with the collar.

QUESTIONS

1. Name and describe the different kinds of fractures.

2. What are the symptoms of fracture?

3. Describe the treatment of fractures.

4. What are the causes of harness injuries?

5. Describe the treatment of the different harness injuries.

CHAPTER XXIII
COMMON SURGICAL OPERATIONS

DEHORNING CATTLE.—It is very often necessary to remove the horns of cattle in order to prevent their injuring or worrying certain individuals in the herd. This operation is of greatest economic importance in dairy and feeding cattle. When first practised, the dehorning of mature cattle was condemned by some persons who deemed it an inhuman and unnecessary operation. It is surely a humane act to remove the horns of cattle that are confined in small yards and pastures, and prevent them from painfully, or seriously, injuring one another.

In most localities there are men who are well equipped to dehorn cattle, and able to perform this operation for a very moderate fee. It is not advisable to attempt to dehorn a number of adult cattle if the operator is not well equipped for the work. Unless a well-constructed dehorning rack is available for confining the animals, there is danger of injuring them and it is very difficult to saw off the horn quickly and satisfactorily. This increases the pain that the animal suffers, and horn stubs soon develop.

Good equipment, such as a chute, saw or clippers, is necessary. A dehorning chute should be built of plank with a good frame well bolted together, with stanchion and nose block for confining the head. Most operators prefer a meat saw for cutting off the horns. It is preferable to dehorning shears, as there is danger of fracturing the frontal bone when removing the horns of mature cattle. The best form of dehorning shears have a wide V in the cutting edge.

The operation is very simple. The horn should be cut off at a point from one-quarter to one-half an inch below the hair line or skin. If this is not practised, an irregular horn growth or stub of horn develops. It is usually unnecessary to apply anything to the wound. If the animal does not strike or rub the part, the clot that forms closes the blood-vessels and the haemorrhage stops. In case of haemorrhage of a serious nature, a small piece of absorbent cotton may be spread over the surface of the wound, and pushed in to the opening in order to keep it in place. Pine tar may be smeared over this dressing. Some operators prefer cauterizing the wound with a red-hot iron for the purpose of preventing haemorrhage. During warm weather, the wound should be washed daily with a two per cent water solution of a coal tar disinfectant, until healing is well advanced. A very necessary after-treatment is the washing of the part after two or three days for the purpose of removing the dried blood.

The opening at the base of the horn communicates directly with the frontal sinus, a large cavity situated between the two plates of the frontal bone. Sometimes the bone is slivered, or the wound becomes infected and inflamed. This may be due to a dirty dehorning saw, or getting dirt into the wound. The inflammation may extend to the sinus and a heavy discharge from the cavity occur. This complication may be prevented by placing the saw or cutters in a disinfectant when not in use, and cleaning and disinfecting the wound very carefully for a few days after the operation.

The horn buttons of calves from a few days to one week of age can be destroyed, and the growth of the horn prevented by applying caustic soda or potash to them. The method of procedure is as follows: Clip away the hair from around the base of the horn tissue and apply a little vaseline to the skin near, but not close to, the base of the horn; moisten the horn button and rub it two or three times with the end of the stick of caustic; do not allow the calf to go out in the rain for a few days after applying the caustic. The horns of calves a few weeks of age may be removed with a sharp knife or calf dehorner.

CHOKING.—This is a common accident in cattle and horses. The object that causes the choke may be lodged in the pharynx or oesophagus. Certain individuals are more prone to choke while feeding than others. This is because of their habit of eating greedily, and swallowing hastily without properly mixing the bolus with the saliva. For this reason, choking occurs when the animal is eating dry feed. Cattle frequently become choked on pieces of such food as roots and apples that are too large to readily pass down the oesophagus. Sharp objects taken in with the food sometimes become lodged in the oesophagus or pharynx.

The symptoms differ in complete and partial choke. In the latter, the symptoms are not very characteristic. The animal may stop feeding, but shows very little evidence of suffering pain. It may be able to swallow a little water. On attempting to drink, a part of the water may be returned through the nose, the same as in complete choke. Ineffectual efforts to swallow, salivation, coughing, hurried respiration, and an anxious expression of the face occur in complete choke. Bloating may complicate this accident in ruminants. After partial choke has persisted for a day or two, the animal appears dejected or distressed. Pressure on the trachea by hard objects may cause difficult respiration.

Mechanical pneumonia sometimes occurs. This is due to the food and water that the animal may attempt to swallow, being returned to the pharynx and passed into the air passages and lungs.

The treatment is as follows: Animals that have choked should not be given access to feed of any kind. Any attempt to take food or drink water may result in pneumonia. It may be necessary to drench the animal with a very small quantity of water for the purpose of diagnosis. The most common form of choke in horses is that due to accumulation of dry food in the oesophagus. The administration of a drug that stimulates the secretion of saliva is a very successful method of relieving this form of choke. Pilocarpine is the drug commonly used. Cheap whips should not be introduced into the oesophagus for the purpose of dislodging the foreign body. There is always danger of the whip becoming broken off, and the broken part lodging in the oesophagus. Neither should such rigid objects as a broom or rake handle be introduced, because of the danger from serious injury to the walls of the pharynx and oesophagus. The flexible probang, which is usually made of spiral wire covered with leather, is a very useful instrument to relieve choke when in the hands of an experienced operator. If the object causing the choke is situated in the neck portion of the oesophagus, it may sometimes be moved forward, or toward the stomach by pressure with the fingers.

CASTRATION.—The castration of the male is a common operation in domestic animals. The purpose of the operation is to render the animal more useful for work or meat production.

The age at which the operation is performed varies in the different species. The colt is usually castrated when he is one year old, and the calf, pig and lamb when a few weeks or a few months of age. It is not advisable to castrate the young at weaning time. The operation and the weaning together may temporarily check the growth of the animal. Colts that are undeveloped and in poor flesh, or affected with colt distemper, should be allowed to recover before they are operated on. In all animals, it is advisable to wait until after they have recovered from disease and become thrifty and strong.

The spring, early summer and fall are the most suitable seasons for castrating the young. It may be practised during the hot or cold months of the year with little danger from wound infection or other complications, providing the necessary after-attention can be given.

The preparation of the animal for the operation by withholding all feed for about twelve hours is very advisable. If this is practised, the stomach and intestines are not distended with feed, and the young are cleaner, easier to handle and suffer less from castration. Clean quarters and surroundings are very necessary to the success of the operation.

The instruments required are sharp knives, preferably a heavy scalpel and a probe-pointed bistoury, an emasculator for large and mature animals, and surgeon's needles and suture material. Ropes and casting harness are frequently used for confining and casting the large and mature animals. Two clean pans or pails filled with a two per cent water solution of liquor cresolis compositus, or an equally reliable disinfectant, should be provided for cleaning the scrotum and neighboring parts and the instruments. Pieces of absorbent cotton or oakum may be used in washing and cleaning the scrotum. The instruments should be sterilized in boiling water before using.

If a number of pigs or lambs are to be castrated, it is best to confine them in a small, clean, well-bedded pen. This enables the attendant to catch them quickly and without unnecessary excitement or exercise. They should be taken to an adjoining pen to be castrated. The scrotum should be washed with the disinfectant, and the testicles pressed tightly against the scrotal wall. An incision parallel with the middle line or raphe and a little to one side is made through the skin and the coverings of the testicle, and the testicle pressed out through the incision. The testicle and cords are then pulled well out and the cord broken off with a quick jerk and twist, or scraped off with a knife. The latter method is to be preferred in large lambs if the operator does not have an emasculator. The incision in the scrotum should be extended from its base to the lowest part, in order to secure perfect drainage.

Young calves may be castrated in the standing position or when cast and held on the side. The method of operating is the same as recommended for pigs and lambs.

The castration of the colt may be performed in either the standing position or when cast. The method of operating is the same as practised in the smaller animals with the exception of cutting off the cord. The emasculator is used here. This instrument crushes the stump of the cord and prevents haemorrhage from the cut ends of the blood-vessels. Careful aseptic precautions must be observed in operating on colts, as they are very susceptible to wound infection and peritonitis.

The blood-vessels of the testicular cord are larger in the adult animals, and the danger from haemorrhage is greater than in the young. For this reason, it is advisable to use an emasculator in castrating all mature animals.

Complications Following Castration.—The *haemorrhage* from the wound and stump of cord is usually unimportant in the young animals. Serious haemorrhage from the vessels of the cord sometimes occur in the adult, and a persistent haemorrhage results when a subcutaneous vein is cut in making the incision in the scrotum. This complication is not usually serious, and can be prevented and controlled by observing proper precautions in cutting off the cord, or by picking up the cut ends of the vessel and ligating it. Packing the scrotal sack with sterile gauze or absorbent cotton, and closing the incision with sutures may be practised for the purpose of stopping this form of haemorrhage. The packing should be removed in about twelve hours.

The infection of the wound always follows castration. If the incision is small and the operation is followed by swelling of the neighboring tissues, the clotted blood, wound secretions and pus become penned up in the scrotal sack. Local blood poisoning or peritonitis follows. This is not an uncommon complication. It can be prevented by aseptic precautions in operating, and insuring good drainage by extending the incision to the lowest part of the scrotal sac. The scrotal sac always contracts down and becomes more or less swollen within a day or two following castration. We must keep this in mind when enlarging the opening, and be sure and make it plenty large to permit the escape of the infectious matter. In castrating sheep, all wool in the region of the scrotal sac should be clipped off, as this interferes with drainage from the wound.

Exercise following castration is almost as essential as clean quarters. Lack of exercise leads to *oedematous swelling* in the region of the scrotum, and the lips of the incision may become adhered if the animal is at rest. Colts and all mature animals that are confined in close quarters should be examined within forty-eight hours following the operation, and the condition of the wound noted. If closed, the hands should be cleaned and disinfected, and the adhesion broken down with the fingers. It is best to exercise horses daily.

It is unsafe to expose castrated animals to cold, damp, chilly weather. The shock and soreness resulting from the operation render the animal highly susceptible to pleurisy and pneumonia. This is especially true of young colts.

Inguinal hernia or *"rupture"* may complicate the operation. This form of hernia is quite frequently met with in pigs, and only occasionally in the other animals. This complication is usually overcome by practising what is commonly termed the covered operation. The pig is usually held or hung up by the hind legs. A larger animal is placed on its back. The hernia is reduced by manipulating the mass of intestines with the fingers, so that they drop back into the abdominal cavity. The part is carefully cleaned and disinfected and an incision made through the scrotal wall, and the thin covering or serous sac in which the testicle is lodged is exposed. The testicle with the cord and covering is drawn well out of the scrotum and held by an attendant. The operator then passes a needle carrying a strong silk thread through the cord and covering, below the point where he intends severing it. The needle is removed and the cord and covering ligated at this point. The cord is then cut off about one-half an inch from the ligature, and the incision in the scrotum made plenty large in order to insure drainage.

It is very essential to the success of this operation that the animal be dieted for twelve or eighteen hours before attempting to operate. The after-treatment consists in giving the animal separate quarters and feeding a light diet.

Enlarged or scirrhous cords follow infection of the wound, usually with spores of a certain fungus (*Botryomyces*). This complication more often follows castration of cattle and pigs than of colts. Wrong methods of operating, such as leaving the stump of the cord too long and insufficient drainage for the pus and wound secretions, are the factors that favor this complication. Scirrhous cords or fibrous tumors should be dissected out and removed before they have become large and begun breaking down.

CASTRATION OF RIDGELING OR CRYPTORCHID ANIMALS.—In the ridgeling animal one or both of the testicles have not descended into the scrotal sac, and are usually lodged in the inguinal canal or abdominal cavity. If the testicle is lodged in the inguinal canal the animal is termed a "flanker." In yearling colts the testicular cord is sometimes short, and the testicle is situated high up in the scrotum and inguinal canal. In examining a supposed cryptorchid colt, he should be twitched. This may cause the testicle to descend into the scrotum.

The castration of a true cryptorchid requires a special operation. When properly performed and the animal given special after-care, the operation is not followed by any serious complications. An abnormally large, diseased testicle is sometimes met with that cannot be removed in the usual way, and which complicates and increases the difficulty of operating.

CAPONIZING.—The castration or caponizing of the male chicken is commonly practised in certain localities. This operation changes the disposition of the cockerel. He becomes more quiet and sluggish, never crows, the head is small, the comb and wattles cease growing and the hackle and saddle feathers become well developed. A capon always develops more uniformly and is larger than the cockerel.

The best time to caponize the cockerel is when he weighs between two or three pounds. If older and heavier, the testicle becomes so large that it is very difficult to remove, and the danger from tearing the spermatic artery and a fatal haemorrhage resulting is greater.

There are several kinds of *caponizing instruments*. They may be purchased in sets. Each set should contain an instrument for removing the testicle; a knife for making the incision through the abdominal wall; a sharp hook for tearing through the thin membrane; spring spreader for holding the lips of the incision apart; a blunt probe for keeping the intestines out of the way of the operator; and a pair of tweezers for removing clots of blood. The different instruments for removing the testicles are a spoon-like scoop, spoon forceps and cannula. The spoon-like scoop is preferred by most operators.

The cockerel is confined for the operation by passing a strong noose of cord around both legs, and a second noose around the wings close to the body, that have weights fastened to them. The cords pass through holes or loops in a barrel or board that is used for an operating table. This holds the cockerel firmly and prevents his struggling.

The bird should be prepared for the operation by withholding all feed and water for a period of twenty-four hours or longer, for the purpose of emptying out the intestine. The operator must have a strong light, in order to work quickly and safely. Direct sunlight or electric light should be used.

The instruments should be placed in a two per cent water solution of carbolic acid. A second vessel containing a two per cent water solution of liquor cresolis compound for cleaning the skin is necessary. Absorbent cotton should be used for washing the wound.

The general method of operating is as follows: The incision is made between the last two ribs and in front of the thigh. The feathers over this region should be removed, and the skin pulled to one side before making the incision. An incision about one and one-half inches in length is made through the skin and muscles, and the spreader inserted. The sharp hook is then inserted and the thin serous membrane over the intestine is torn through. The testicles are situated in the superior portion of the abdominal cavity or under the back. On pushing the intestines to one side, both testicles, which are about the size of a bean and yellowish in color, can be seen. The lower one should be removed first. After removing both testicles, blood clots, feathers, or any foreign body that may have gotten into the wound should be picked up with the tweezers before removing the spreaders and allowing the wound to close. No special after-treatment is required.

The most common complication is rupture of the spermatic artery. This occurs at the time the testicle is torn loose and may be due to careless methods, or operating on cockerels that are too large. If all of the testicle is not removed from the abdominal cavity, the bird is termed a "slip."

Sometimes air puffs form after the operation. These should be punctured with a sharp knife.

OVARIOTOMY, "SPAYING."—The removal of the ovaries, or ovariotomy, is practised for the purpose of rendering the female more useful for meat production, prolonging the period of lactation, overcoming vicious habits and preventing oestrum or heat. The operation is commonly performed in the heifer and bitch, occasionally in the mare, and at present rarely in the sow.

Heifers are usually spayed between the ages of eight and twelve months; the *bitch* and *sow* when a few months old, or before the periods of heat have begun. The *mare* is spayed when mature. It is possible to spay the female at any age, but the ages mentioned are the most convenient. Pregnant animals should not be operated on. The season of the year makes little difference in the results, providing the animal can be kept under close observation and given the necessary care and treatment. The spring of the year, just before turning the herd on pasture, is the best season to spay heifers.

All animals should be prepared for the operation by withholding all feed for at least twenty-four hours before they are operated on, and it may be advisable to give them a physic. It is easier to operate when the intestinal tract is comparatively empty, and the death rate is lower than when the animal is not properly prepared for the operation.

The method of operating is not the same in the different species. In young heifers and sows, the flank operation is preferred, and in mares and cows, the vaginal operation. The median line operation is practised in bitches. A spaying emasculator, or ecraseur, are the special instruments need for removing the ovaries.

The animal must be properly confined for the operation. Heifers are usually held in the standing position by fastening the head securely, and crowding the left side of the animal against a solid board partition, or side of a chute. If the vaginal operation is performed, the mare or cow may be confined in stocks. The bitch is usually anesthetized and placed on her back on a table that is inclined, so that the hind parts are elevated.

Ovariotomy cannot be successfully performed by an untrained and inexperienced operator. The necessary precautions against the infection of the part must be observed, in order to promote the healing of the wound and prevent peritonitis. The seat of the operation should be carefully cleaned and disinfected.

Following the operation the animal should be fed a spare diet for a few days. This is a very necessary part of the care of the bitch. The general condition of the animal should be noted daily until there is no further danger from wound infection. Healing is usually completed in from seven to twelve days. The sutures should then be removed, and if stitch abscesses occur, the part should be washed with a disinfectant.

QUESTIONS

1. What is the purpose of dehorning cattle? Give different methods of removing the horns.

2. Give the causes and treatment of choking.

3. What is the purpose of castration and ovariotomy?

4. At what age is it best to practise castration and ovariotomy?

5. In what way should an animal be prepared for castration? Give a description of the method of castration in the different animals.

6. What special care should be given following castration?

7. What are some of the complications that may follow castration?

PART V
PARASITIC DISEASES

CHAPTER XXIV
PARASITIC INSECTS AND MITES

Parasitic insects are common causes of skin diseases in domestic animals. The diseased conditions of the skin, and the irritation that they may cause the animal, depend on the life history and habits of the parasite. Species that are unable to live independently of a host and are permanent parasites are usually the most injurious to the animal. This is especially true of parasites that are capable of puncturing the skin or burrowing into it. Temporary parasites may cause fatal forms of disease. This is true of the larva? of the sheep bot-fly, which develop in the sinuses of the head, causing severe inflammation of these parts, nervous symptoms and death. The character of the symptoms of a parasitic disease depends on the habits of the parasite, and the tissue or organ, that it may attack.

The parasitic flies belong to the order *Diptera*, and the families *Muscidae* and *OEstridae*. Fleas belong to the sub-order *Pulicidae*. The order *Hemiptera* includes the lice, and the most important families are *Pediculidae* and *Ricinidae*. Mites and ticks belong to the order *Acarina*. The most important parasites belonging to this order are the *Sarcoptidae* and *Ixodidae*.

OESTRIDAE.—The three common bot-flies are the *Gastrophilus equi*, *Hypoderma lineata* and *OEstrus ovis*. These flies are important because of the parasitic habits of their larva. They inhabit the stomach and intestines of horses ; the subcutaneous tissue and skin of cattle; and the sinuses of the head and nasal cavities of sheep.

The common bot-fly of the horse (*G. equi*) has a heavy, hairy body. Its color is brown, with dark and yellowish spots. The female fly can be seen during the warm weather, hovering around the horse, and darting toward the animal for the purpose of depositing the egg. The color of the egg is yellow, and it adheres firmly to the hair. It hatches in from two to four weeks, and the larva reaches the mouth through the animal licking the part. From the mouth, it passes to the stomach, where it attaches itself to the gastric mucous membrane . Here it remains until fully developed, when it becomes detached and is passed out with the fasces. The third stage is passed in the ground. This takes place in the spring and early summer and lasts for several weeks, when it finally emerges a mature fly.

The bot-fly of the ox (*H. lineata*) is dark in color and about the size of a honey-bee. On warm days, the female may be seen depositing eggs on the body of the animal, especially in the region of the heels. This seems to greatly annoy the animal, and it is not uncommon for cattle to become stampeded. The egg reaches the mouth through the animal licking the part. The saliva dissolves the shell of the egg and the larva is freed. It then migrates from the gullet, wanders about in the tissue until finally it may reach a point beneath the skin of the back. Here the larva matures and forms the well-known swelling or warble. In the spring of the year it works out through the skin. The next stage is spent in the ground. The pupa state lasts several weeks, when the mature fly issues forth.

The bot-fly of sheep (*O. ovis*) resembles an overgrown house-fly. Its general color is brown, and it is apparently lazy, flying about very little. This bot-fly makes its appearance when the warm weather begins, and deposits live larvae in the nostrils of sheep. This act is greatly feared by the animals, as shown by their crowding together and holding the head down. The larva works up the nasal cavities and reaches the sinuses of the head, where it becomes attached to the lining mucous membrane. In the spring, when fully developed, it passes out through the nasal cavities and nostrils, drops to the ground, buries itself, and in from four to six weeks develops into the mature fly.

SYMPTOMS OF BOT-FLY DISEASES.—The larvae of the bot-fly of the horse do not cause characteristic symptoms of disease. Work horses that are groomed daily are not hosts for a large number of "bots," but young and old horses that are kept in a pasture or lot and seldom groomed may become unthrifty and "pot bellied," or show symptoms of indigestion.

Cattle suffer much pain from the development of the larva of the *H. lineata*. During the spring of the year, the pain resulting from the presence of the larvae beneath the skin and the penetration of the skin is manifested by excitement and running about. Besides the loss in milk and beef production, there is a heavy yearly loss from the damage to hides.

The parasitic life of the bot-fly of sheep results in a severe catarrhal inflammation of the mucous membrane lining the sinuses of the head, and a discharge of a heavy, pus-like material from the nostrils. The irritation produced by the larvae may be so serious at times as to result in nervous symptoms and death.

TREATMENT OF BOT-FLY DISEASES.—The treatment of the different bot-fly diseases is largely preventive. This consists in either the destruction of the eggs or the larvae.

The different methods of destroying the eggs of the bot-fly of the horse are clipping the hair from the part, scraping off the eggs with a sharp knife, or destroying them by washing the part infested with eggs with a two or three per cent water solution of carbolic acid. This should be practised every two weeks during the period when the female deposits the eggs.

Housing the cattle, or applying water solutions of certain preparations to the skin that may keep the female from depositing eggs, may be practised for the prevention of the ox-warble. The most practical method of ridding cattle of this pest is to destroy the larvae. This can be done by examining each animal and locating the swelling or warble and injecting a few drops of kerosene into the opening in the skin. A better method is to enlarge the opening in the skin with a sharp knife, squeeze out the grub and destroy it. This should be practised in late winter and early spring.

The application of pine tar to the nostrils of sheep is the most practical method of preventing "grub in the head." This should be practised every few days during the summer months. A very good preventive measure is plenty of shade for the flock. Valuable animals may be treated by trephining into the head sinus and removing the "grub."

LICE.—The sucking lice belong to the genus *Hoematopinus*, and the biting lice of mammals belong to the genus *Trichodectes*. Different species of sucking and biting lice occur on the different species of farm animals. Poultry act as hosts for many different species of biting lice belonging to the following genuses: *Lipiurus, Goniodes, Goniocotes* and *Menopon*.

The common sucking lice occurring on animals are the large-headed horse louse, *H. macrocephalus*; the long-nosed ox louse, *H. tenuirostris*; the large-bellied ox louse, *H. curysternus*; the *H. stenopses* of sheep; *H. suis* of swine; and the *H. piliferus* of the dog.

The *common biting lice* that are found on domestic animals are the *T. pilosus* and *T. pubescens* of solipeds, *T. scalaris* of the ox, *T. spoerocephalus* of sheep and goats, *T. latus* and *T. subrostratus* of the dog and cat. *Menopon palidum, Lipiurus variabilis* and *Gonoides dissimilis* are the common lice found on poultry.

SYMPTOMS OF LICE.—The symptoms of lousiness depend on the variety of lice present, the degree to which the animal is infested with them, its physical condition and the care that it receives. Lice multiply more rapidly and cause greater loss during the winter months than they do in the summer, when the animals are not housed and the opportunity for infection from the surroundings is not so great. The sucking louse is the most injurious and irritating. The irritation and loss of blood that the animal may suffer when badly infested by this parasite may result in marked unthriftiness. Young and old animals that are not well cared for suffer most. The biting louse may bite through the superficial layer of the skin, and cause the animal to bite and rub the part. This irritation to the skin prevents the animal from becoming rested, and after a time seriously interferes with its thriftiness.

Horses and mules show a staring, dirty, rough coat. The mane and tail may become broken and matted. The animal rubs against the stall, fences and trees, and bites the skin in its efforts to relieve the irritation. On examining the coat, nits are found adhering to the hair . We should examine the parts of the skin covered by the long hair for the sucking lice; and the withers, abdomen and limbs for the biting lice.

The symptoms of lousiness in cattle are about the same as occur in horses. Licking and rubbing the skin are prominent symptoms in cattle, and the coat becomes dirty and rough. The licked part is matted and curled. The lice may be discovered by parting the hair along the back and rump.

The biting louse of sheep causes the fleece to become matted and tufts of wool are pulled out. This is brought about by the sheep rubbing and nibbling the fleece, and the lice cutting through the wool. The loss due to the damage to the fleece is usually greater than that resulting from unthriftiness.

The hog-louse is the largest specie known. As well as the largest, it is the most common of all lice found on domestic animals. The favorite points of attack are the under surface of the body, the neck and the inside of the thighs. The irritation and itching are severe, and the hog rubs and scratches the skin. Young hogs suffer most from this parasite, and their thriftiness is greatly interfered with.

The long-haired breeds of *dogs* suffer more from lice than the short-haired breeds. The almost constant scratching and biting of the skin result in its becoming badly irritated and scabby. The symptoms differ little from irritation to the skin caused by fleas, but the presence of biting or sucking lice enables the person making the examination to determine the cause of the irritation.

Lice are the most common parasites of *poultry*. It is uncommon to meet with a flock of fowls that are not hosts for one or more of the many different varieties of bird lice. Restlessness, picking, scratching, flapping the wings, abandoning the nest and loss of condition are common symptoms. Young birds suffer most from lice. This is especially true of young chickens, death frequently resulting. Old fowls may show little inconvenience unless badly infested. The finding of the lice with the head imbedded in the skin or on the feathers enables the person making the examination to positively diagnose the case. The head, back, region of the vent and beneath the wings are the parts that should be carefully examined for lice.

TREATMENT OF LOUSINESS.—The preventive treatment is very important. This consists in carefully examining all animals or birds that have been purchased recently, and if found to harbor lice, excluding them from the herd or flock until after they have been properly treated.

It is impossible to rid animals of lice if the quarters are thoroughly cleaned and disinfected. This is necessary in order to destroy lice that have become scattered about by the lousy animals, and prevent the reinfection of the treated animals. The best method to use in cleaning the quarters is to remove all litter and manure from the stable or houses and their immediate surroundings. It should be burned, or hauled to a field or lot where other animals cannot come in contact with it for a few months. The walls, floors and partitions should be sprayed with a three per cent water solution of liquor cresolis compositus. Lime may be scattered about the buildings, yards and runs. The most satisfactory method of destroying lice on the bodies of animals is by washing or dipping in a water solution or mixture of some reliable disinfectant or oil.

Running hogs through a dipping tank that contains a one or two per cent water solution of liquor cresolis compositus, or a coal tar disinfectant, or that has from three-fourths to one and one-half inches of oil on top of the water, is the most satisfactory method of destroying the hog louse. Because of the thinness of the hog's coat and the danger from irritating the skin when strong solutions of a disinfectant are used, most swine breeders prefer crude oil as a remedy for lousiness in hogs. Crude oil may be applied to the bodies of hogs with a swab. If this method is practised instead of dipping, it is advisable to crowd the hogs into a small pen, and apply the oil in front and between the thighs and back of the arms. This may be practised during the cold weather when it is impossible to dip the animals.

Horses may be washed with a one or two per cent water solution of liquor cresolis compositus, or a coal tar disinfectant. If the weather is cold, it is advisable to pick a sunny day, and blanket the animal after rubbing it as dry as possible in order to prevent chilling and catching cold.

Cattle may be treated in the same manner as horses. Mercurial ointment rubbed in small amounts on the skin back of the horns and ears, where the animal cannot lick it, is a common remedy. The absorption of a small amount of this drug does the animal no harm, but a larger quantity may salivate it.

Sheep are treated by dipping in a water solution of a reliable coal tar disinfectant. This should not be practised during cold weather, as the fleece does not dry out. Insect powder may be dusted into the fleece when it is impossible to dip the animal.

A very satisfactory treatment for lousiness in *dogs* and *cats* is to wash them with carbolized soap. We should wait a few minutes before rinsing off the soapy lather and drying the coat.

A number of different remedies are used for the treatment of lousiness in *poultry*. Dust baths and insect powder are recommended. Ointments are commonly used. One part sulfur and four parts vaseline, or lard, may be made into an ointment and applied to the head, neck, under the wings and around the vent. Mercurial ointment may be applied to the margin of the vent. Neither of them should be used for destroying lice on young chicks. Mercurial ointment should be used very carefully because of its poisonous effect. Lard may be used for destroying lice on young chicks. Crude petroleum may be sprayed among the feathers by a hand-sprayer, while the fowls are suspended by the feet.

None of the disinfectants and oils recommended for dipping and washing lousy animals destroy the nits. This makes it necessary to re-treat the animal in from eight to ten days after the first treatment.

THE SHEEP-TICK.—This is not a true tick. It resembles a fly more than it does a tick, and its right name is *Melophagus ovinus* . Louse-fly is a better name for this parasite than tick, as its entire life is spent on the body of a sheep. The general color of the body is brown. The legs are stout, covered with hair and armed with hooks at their extremities. The mouth parts consist of a tubular, toothed proboscis with which the parasite punctures the skin and sucks the blood. Within a few hours after birth, the larvae develop into pupae, which are hard, dark brown in color and firmly glued to the wool. The young louse-fly emerges from the pupa in from three to four weeks.

The sheep-tick is a very common external parasite. The adult parasites and the pupae are large and easily found. When badly infested with ticks, a sheep will rub, dig and scratch the skin and fleece. This results in pieces of wool becoming pulled out and the fleece appears ragged. After clipping the ticks migrate from the ewes to the lambs, which may become unthrifty and weak.

The treatment consists in dipping the flock in a one or two per cent water solution of a coal-tar dip. Dips containing arsenic are most effective in ridding sheep of ticks.

SCABIES. — This parasitic disease is one of the oldest and most prevalent diseases of the skin. It is commonly known as scab or mange. The animals most commonly affected are sheep, horses and cattle.

The disease is caused by *small mites* or *acari* that are naturally divided into the *Sarcoptes*, which burrow under the epidermis, forming galleries; the *Psoroptes*, which live on the surface of the skin where they are sheltered by scabs and scurf; and the *Symbiotes*, which also live on the surface of the skin, but prefer the regions of the hind feet and legs.

Acari multiply rapidly and live their entire life on the body of the host. A new generation is produced in about fifteen days. Gerlach has estimated the natural increase in three months at 1,000,000 females and 500,000 males. Scab and mange are exceedingly contagious diseases.

Common sheep scab is caused by that specie of mites known as the *Psoroptes communis var. ovis* . Any part of the body may become affected. The bites of the mites greatly irritate the skin, and the animal scratches, bites and rubs the part in its effort to relieve the intense itching. The skin becomes inflamed and scabby, the wool is pulled and rubbed out, and the fleece becomes ragged . By pulling wool out of the newly infested area, or collecting skin scrapings and placing this material on black paper in a sunny, warm place, the mites may be seen crawling over the paper. This method of diagnosis should be resorted to in all suspicious cases of skin disease, and before the disease has developed to any great extent.

The mite that most commonly causes *mange in cattle* is the *Psoroptes communis var. bovis.* It may invade the skin in the different regions of the body, but it is in the regions of the tail and thighs that the first evidence of the mange is noticed. The animal rubs, scratches, and licks the part. The itching is intense. The hair over the part is lost and the skin appears inflamed, thickened, moist, or covered with white crusts. Cracks and sores may form in the skin. The examination of scrapings from the inflamed skin should be practised in order to confirm the diagnosis.

Mange in horses may be caused by either psoroptic or sarcoptic mites. *Psoroptes communis var. equi* seems to be the more common parasite. The itching is intense. The inflamed areas are small at first and scattered over the regions of the rump, back and neck . After a time the small areas come together and form large patches, and further spreading of the inflammation results from grooming, scratching and biting the skin. Scattered, elevated eruptions on the skin from which the hair has dropped out are first noticed.

These parts may show yellowish scabs. Later the skin is thickened, smooth, wrinkled, cracked, or covered with sores. Scrapings made from the inflamed areas of the skin may show the psoroptic mites.

Mange in hogs is comparatively rare. It is caused by one of the sarcoptic mites. The thin portions of the skin are usually first invaded. There are violent itching and rubbing, and small, red elevations occur on the skin in the region of the ears, eyelids or inner surface of the thighs, depending on the part first invaded. The skin becomes greatly thickened and covered with crusts and scabs. Pus formation and ulceration may occur.

TREATMENT OF SCAB AND MANGE.—A careful inspection of recently purchased animals that pass through stockyards, or are shipped from sections where scab and mange are common skin diseases, is an important preventive measure. Infected animals should be completely isolated from the herd, and kept apart from other animals until after they have been treated. Hogs that are slightly infected should be quarantined and treated. If badly affected, they should be killed, and the carcass disposed of by burning or burying.

The different remedies used in the treatment of the disease may be applied by dipping, hand dressing or washing, pouring, smearing and spotting. The first method is the most satisfactory. The last method may be used when a small area of the skin is involved, and during the cold weather. Washing or dipping the animal with a two per cent water solution of liquor cresolis compositus is an effective remedy for the psoroptic forms of scab and mange. Tobacco, lime and sulfur, and arsenical dips are recommended in the treatment of sheep and cattle. Ointments are recommended for animals that are slightly affected with mange. Lime and sulfur dips are recommended by the Bureau of Animal Industry. Small infected areas of the skin may be treated by applying sulfur-iodide ointment. The following ointment is commonly recommended: potassium sulfide ten parts, potassium carbonate two parts, and lard three hundred parts.

Sheep cannot be safely dipped for scab during the cold weather. If thickened and scabby, the skin should be scrubbed with the dip, or the animal prepared for dipping or washing by first clipping the hair or wool and scrubbing the skin with water and a good soap. In order to prevent reinfection, it is necessary to remove the animal to new quarters, or thoroughly clean and disinfect the old. It is necessary to wash or spray the fences, floors, walls, brushes and curry-combs with a disinfecting solution.

Manure and other litter should be removed to a place where there is no danger from its distributing the infection.

DISEASES OF POULTRY CAUSED BY MITES.—Mites or acarina that cause diseases of poultry may live on the feathers, beneath the skin, and within the body of the fowl.

The small, red mite (Dermanyssus gallinae) remains on the surface of the body only when feeding, and spends the rest of the time under collections of filth and in cracks in the roosts and walls of the house. This parasite causes the birds to become restless, emaciated and droopy.

A very small mite (Sarcoptes mutans) is the cause of scaly leg. It lives under the skin. The joints of the feet appear affected, and the foot and leg become enlarged, roughened and scaly.

Depluming scabies is caused by *Sarcoptes laevis var. gallinae.* This mite causes the feathers to break off at the surface of the skin. Masses of epidermic scales may form around the broken ends of the feathers. The diagnosis can be confirmed by examining the skin lesions and finding the mite.

The air sac mite (Cytodites nudus) may cause sufficient irritation to the mucous membrane lining the air sacs to seriously obstruct the air passages with mucus, or produce death from exhaustion. A post-mortem examination of a fowl that has died of this disease shows the mites on the surface of the lining membrane of the air-sacs. They appear as a white or yellow dust.

TREATMENT OF POULTRY DISEASES CAUSED BY MITES.—Diseases of poultry caused by mites may be prevented by quarantining all recently purchased birds for a period of from two to four weeks, and by keeping the poultry houses clean. Birds that are found infested with parasites should be destroyed or returned. In case the bird is valuable and suffering from external parasites only, it should be given the necessary treatment.

Red mites may be destroyed by thoroughly cleaning the poultry house, and spraying the roosts, nests, walls and floor with a three per cent water solution of liquor cresolis compositus. This should be repeated twice a week for two weeks.

Scaly-leg may be treated by applying a penetrating oil to the feet and lower part of the leg. It is advisable to first remove the scales by scrubbing the part with soap and warm water. Dipping the feet in a mixture of kerosene one part and linseed oil two parts is recommended. This should be repeated as often as necessary.

QUESTIONS

1. Describe the different bot-flies.

2. Give the life history of the bot-fly of the horse; of the ox; of sheep.

3. Give the symptoms of bot-fly diseases.

4. Give the symptoms of lousiness.

5. Give treatment for lousiness of different farm animals.

6. What is the damage from the sheep-tick? Give treatment.

7. Describe the injury from scabies and mange.

8. Give treatments for these diseases.

9. Mention the several poultry mites and tell how to treat them.

CHAPTER XXV
ANIMAL PARASITES

The common parasitic diseases of domestic animals are caused by the following groups of worms: *Flukes* or *trematoides*; *tapeworms* or *Cestoides*; *thorn-headed worms* or *Acanthocephales*; and *round-worms* or *Nematoids*. Flat worms, such as tapeworms and flukes, require secondary hosts. The immature and mature forms of tapeworms are parasites of vertebrate animals, but an invertebrate host is necessary for the completion of the life cycle of the fluke. The hog is the only specie of domestic animals that becomes a host for the thorn-headed worm. The round-worm is a very common parasite. There are many species belonging to this class.

DISTOMA HEPATICUM (COMMON LIVER FLUKE).—Sheep are the most common hosts for this parasite. It is present in the gall ducts and livers, and causes a disease of the liver known as liver rot. The liver fluke is flat or leaf-like and from thirteen to fifteen mm. long . The head portion is conical. It has an oval and ventral sucker, and the body is covered with scaly spines. The eggs are oval and brownish in color.

The life history, in brief, is as follows: Each adult is capable of producing an immense number of eggs which are carried down the bile ducts with the bile to the intestine, and are passed off with the faeces. Under favorable conditions for incubation, such as warm, moist surroundings, the ova or eggs hatch and the *ciliated embryos* become freed. The embryo next penetrates into the body of certain snails and encysts. The *sporocyst*, as it is now called, develops into a third generation known as *redia* which escape from the cyst. The *daughter redia* or *cercaria*, as they are now termed, leave the body of the snail and finally become encysted on the stems of grass, cresses and weeds. When taken into the digestive tract of the animal grazing over infested ground, the immature flukes are freed by the digestive juices. They then pass from the intestine into the bile ducts. The period of development varies from ten to twenty weeks; each sporocyst may give rise to from five to eight *redia* and each redia to from twelve to twenty *cercaria*.

Fluke diseases occur among animals pastured on low, wet, undrained land. Drying ponds and lakes are the homes of the fresh water snails, and in such places there are plenty of hosts for the immature flukes. Wet seasons favor the development of this parasite. Cattle and sheep that pasture on river bottom land in certain sections of the southern portion of the United States are frequently affected with fluke diseases.

The symptoms of liver rot of sheep may be divided into two stages. The first stage is marked by increase in weight and improved condition. In the second stage of the disease, the animal shows a pale skin and mucous membrane, dropsical swellings, loss of flesh and weakness. The character of the symptoms of the disease depends on the age of the animals and the care that they receive. Young, poorly cared for animals suffer severely from the disease, and the death rate is usually heavy. The finding of fluke ova in the faeces is conclusive evidence of the nature of the disease. It may be advisable to kill one of the sick animals, and determine the nature of the disease by a post-mortem examination.

The treatment is preventive. Drainage water from a pasture infested with snails harboring immature flukes is a source of infection, and should not be used as a water supply for cattle and sheep. In sections where the disease is prevalent, sheep should not be pastured on low, poorly-drained land. Such land should be used for pasturing horses and cattle, but if possible, it should be first drained and cultivated. Careful feeding and good care may help the affected animals to recover.

TAPEWORMS OR CESTOIDES.—Tapeworms are formed by a chain of segments, joined together at their ends, and are flat or ribbon-shaped . The head segment is small, and possesses either hooks or suckers. It is by these that the worm attaches itself to the lining membrane of the intestine. The anterior segments are smaller and less mature than the posterior segments. Each segment is sexually complete, possessing both the male and female organs, and when mature, one or more of them break off and are passed out with the faeces. The mature or ripe segments are filled with ova. On reaching the digestive tract of a proper host, usually with the drinking water or fodder, the embryo is freed from the egg. The *armed embryo* uses its hooklets in boring its way through the wall of the intestine. It then wanders through the tissues of its host until it finally reaches a suitable place for development (Figs. 71 and 73). On coming to rest, it develops into the larva or bladder-worm, which when eaten by a proper host gives rise to the mature tapeworm.

The following tables give the most important tapeworms:

ADULT FORMS

Name Host Organ

Taenia expansa Sheep and ox Intestine
Taenia fimbriata Sheep Liver
Taenia denticulata Cattle Intestine
Taenia alba Cattle Intestine
Taenia perfoliata Solipeds Intestine
Taenia mamillana Solipeds Intestine
Taenia echinococcus Dog Intestine

LARVAL FORMS

Name Host

Cysticercus bovis Cattle
Cysticercus cellulosa Swine and man
Cysticercus tennicollis Cattle, sheep and swine
Coenurus cerebralis Cattle and sheep
Echinococcus polymorphus Cattle, sheep, swine and man

The adult tapeworms *Taeniae saginata* and *soleum*, of which the *Cysticerci bovis* and *cellulosa* are the larvae forms, occur in man. The larvae are present in meat and pork, and this form of parasitism is termed beef measles in cattle and pork measles in hogs. Man becomes host for these two forms of tapeworms through eating measly pork or beef that is not properly cooked.

The dog is the host for *Taeniae marginala, coenurus* and *echinococcus*. The larvae forms of these *taeniae* are the *Cysticercus tennicollis, Coenurus cerebralis* and *Echinococcus polymorphus. C. tennicollis* is a parasite of the serous or lining membranes of the body cavities. It is not of great economic importance. *C. cerebralis* is a parasite of the brain of sheep, and may cause a heavy death rate in flocks that are infested with it. *E. polymorphus* is a parasite of the liver, but it may occur in other organs.

THE THORN-HEADED WORM OR ACANTHOCEPHALE.—This parasite requires a secondary host. In this case a particular species of the May-beetle larva or white grub that is commonly found about manure piles and in clover pastures is the host. The hog eats a white grub that is host for the larval form. The digestive juices free the larva, it then becomes attached to the intestinal mucous membrane and develops into the adult thorn-headed worm . This parasite is characterized by a hooked proboscis or thorn at its anterior extremity, and the absence of a distinct digestive tract. The male is much smaller than the female. The eggs are passed out of the intestine with the faeces.

THE ROUND-WORMS OR NEMATOIDS.—Round-worms are very common parasites of domestic animals . This group of worms is characterized by their cylindrical form, the presence of a true digestive canal and the separation into two sexes, male and female. The life history is more simple than in the flat worms. Intermediate hosts are not required for the development of the common forms. The eggs and embryos are deposited by the female in the intestinal tract, air passages, or excretory ducts of the kidneys of the host. Development may be completed here, or the eggs and embryos are passed off with the body excretions. They may live for a short time outside the animal body, or undergo certain development and again infest a host of the same species from which they came, through the water, grass and fodder that the animal may take into its digestive tract.

The following species of nematoids are common parasites of domestic animals:

SOLIPEDS

Species Organ
Ascaris megalocephala Intestines
Sclerostoma equinum Large intestine and blood-vessels
Sclerostoma tetracanthum Large intestine
Oxyrus curvula Large intestine
Oxyrus mastigodes Large intestine

CATTLE

Species Organ
Strongylus convolutus Abomasum
Ascaris vituli Small intestine (calves)
Strongylus ventricosus Small intestine
Oesophagostomum inflatum Large intestine
Trichocephalus affins Large intestine
Strongylus micrurus Bronchi
Strongylus pulmonaris Bronchi

SHEEP

Species Organ
Haemonchus contortus Abomasum
Ascaris ovis Small intestine

Strongylus filicollis Small intestine
Oesophagostomum columbianum Intestines
Uncinaria cernua Small intestine
Trichocephalus affins Large intestine
Strongylus filaria Bronchi
Strongylus rufescens Bronchi and air follicles

SWINE

Species Organ
Ascaris suis Intestines
Oesophagostomum dentatum Large intestine
Trichocephalus crenatus Large intestine
Trichina spiralis Muscles and intestines
Strongylus paradoxus Trachea and bronchi
Sclerostoma pingencola Renal fat and kidney

POULTRY

Species Organ
Ascaris inflexa Intestine
Spiroptera hamulosa Gizzard
Heterakis papillosa Caecum
Syngamus trachealis Trachea and bronchi

INTESTINAL WORMS OF SOLIPEDS.—The large round-worms or ascarides and the sclerostomes are the most injurious intestinal parasites of solipeds. The *A. megalocephala* or large round-worm is from 5 to 15 inches (12 to 35 cm.) long. It may be present in the double colon in such large numbers as to form an entangled mass that completely fills a portion of the loop in which it is lodged. It may interfere with digestion by obstructing the passage of alimentary matter, and irritating the intestine.

The *S. equinum* and *S. tetracanthum* are small worms. The former sclerostoma is from 0.6 to 1.5 inches (18 to 35 mm.) long, and the latter is from 0.5 to 0.6 inch (8 to 17 mm.) long. Both sclerostomes attach themselves to the lining membrane of the intestine by their mouth parts, and suck blood. The young *S. equinum* may live in tumor-like cysts that they cause to form in the lining membrane of the intestine. The young worm may penetrate the wall of a small blood-vessel as well, and drift into a large vessel, where it may become lodged and undergo partial development. The irritation to the blood-vessel results in an inflammation and dilation of the vessel wall. This is termed verminous aneurism. A portion of the fibrin-like lining of the aneurism may flake off and drift along in the blood stream, until finally a vessel that is too small for the floating particle or embolus to pass through is reached. The vessel is then plugged or a thrombus is formed. If the vessel involved by the thrombus happens to be a mesenteric vessel, then a loop of intestine has its blood supply cut off, and colicky pains result. Such colics are dangerous, and may terminate fatally. Intestinal obstruction, thrombo-embolic colics, unthriftiness and a weakened, anaemic condition may be caused by intestinal worms.

The treatment is both preventive and medicinal. The preventive treatment consists in giving young, growing animals the best care possible. Cleanliness about the stable, giving the colt plenty of range when running in a pasture, and feeding a ration that is sufficient to keep the colt in good physical condition are the important preventive measures. Tartar emetic in one-half to one dram doses may be given with the feed daily until five or six doses are given. Turpentine may be given in one to three ounce doses in a pint of linseed oil. This may be repeated daily for two or three days. Worms located in the posterior bowel may be removed by rectal injections of a weak water infusion of quassia chips. The rectum should be first emptied with the hand, and the nozzle of the syringe carried as far forward with the hand as possible. The injections should be repeated daily for several days.

INTESTINAL WORMS OF CATTLE.—Intestinal worms seldom cause serious losses from unthriftiness or death in cattle. It is in calves only that we are called on to treat this class of disease. The symptoms resulting from the invasion of the intestinal tract by the different worms vary in severity according to the number, habits of the parasite and care that the animal receives. The usual symptoms are unthriftiness, indigestion, diarrhoea and a stunted, anaemic condition. Stiles reported extreme anaemia, unthriftiness and many deaths among cattle in a certain section of Texas, due to extensive infection with the *Uncinaria radialus*.

The treatment is largely preventive. Calves and yearlings should be provided with plenty of feed at all seasons of the year. Good care and careful feeding will keep them in a thrifty, healthy condition and enable them to throw off invasions of intestinal worms. Turpentine is the vermifuge usually administered to calves. The dose is from two to four drams given in a milk or raw linseed oil emulsion.

STOMACH WORM OF SHEEP.—The twisted stomach worm, *Haemonchus contortus*, is the most injurious internal parasite of sheep. It is a very small, hair-like worm from 0.4 to 1 inch (9 to 25 mm.) in length. In the adult form it attaches itself to the mucous membrane of the fourth stomach or abomasum, and lives by sucking blood. The blood present in the digestive tract of the worm gives it a brown color, and the white oviducts which are wound around the digestive canal cause the body to appear twisted. When the twisted stomach worm is present in large numbers, the worms become mixed with the contents of the stomach and can be readily found on making a post-mortem examination.

Symptoms of stomach worms are first manifest in the lambs . It is not until early summer that the disease appears in the flock. The symptoms are not characteristic unless we consider an unthrifty, anaemic, weak, emaciated condition accompanied by diarrhoea, during the summer months characteristic of stomach-worm disease. The sick animals are unable to keep up with the flock, and they like to stand about in the shade. They move slowly, the back is arched, the appetite poor, the mucous membranes and skin are pale and the hind parts soiled by the diarrhoeal discharge. More acute symptoms than the above sometimes occur. The disease may last from a few days to several weeks. A large percentage of the affected animals die.

The *treatment* is largely preventive. Frequent changing of pastures and dry lot feeding are common preventive measures. Permanent sheep pastures lead to heavy losses from stomach worm disease. A very effective preventive measure, as we may term it, is the practice of administering a vermifuge to the ewes in the late summer and again in early winter. This may be given in a drench, or with the feed. This prevents the reinfection of the pastures every spring, and the young lambs are not exposed to this form of infection. The most effective treatment that the writer has ever used is the following formula recommended by Dr. Law: Arsenous acid one dram, sulfate of iron five drams, powdered areca nut two ounces, common salt four ounces. This is sufficient for one dose for thirty sheep. It may be given with the salt, or in ground feed. If the flock is apparently healthy, four doses given at intervals of three days is sufficient. If symptoms of stomach worms are manifested the animals should be dosed daily until they have received from five to ten doses, depending on the condition of the animal.

INTESTINAL WORMS OF SHEEP.—The most widely distributed and seemingly most injurious intestinal worm of sheep is the *OEsopliagostomum columbianum*. It is a small worm from 0.5 to 0.75 inch (12 to 18 mm.) long. It penetrates the lining membrane of the intestines and encysts in the intestinal wall. A tumor, varying in size from that of a millet seed to a hazelnut, then forms in the wall of the intestine. These tumors undergo a cheesy degeneration, and when mature, may appear as greenish, cheesy-like masses, covering a large portion of the lining membrane of the intestine. Diarrhoea and emaciation may result. These symptoms are most evident during the winter months.

The treatment recommended for ridding sheep of this intestinal worm is largely preventive. Very little can be done with the medicinal treatment of a sheep whose intestinal tract is badly infested with this parasite. Good care and the feeding of a proper ration are the only curative measures that are effective in such cases. The occasional administration of a vermifuge for the purpose of ridding the digestive tract of worms, together with the frequent changing of pastures during the spring and summer, are the most effective preventive lines of treatment. The same treatment recommended for stomach worms may be used for this disease.

INTESTINAL PARASITES OF HOGS.—The *Ascaris suis* or *common round* worm is very commonly found in the small intestine. It is quite frequently found in large numbers, almost filling the lumen of the intestine of an unthrifty pig . It may also work its way into the bile duct. Sometimes, after a hog has died, this parasite migrates forward into the stomach and gullet. The *A. suis* is from 4 to 10 inches (10 to 26 cm.) long.

The Echinorhynchus gigas or *thorn-headed* worm is the most dangerous of all intestinal worms . It is usually found with its proboscis or thorn imbedded in the wall of the small intestine. The Echinorhynchus is not as common a parasite as the Ascaride, and it is not usually present in large numbers. Usually, not more than a half-dozen of these worms are found in the intestine of a hog, but in some localities and in hogs that are allowed to root around manure piles and in clover pastures the herd may become badly infected with them and serious losses occur. The average length of the male is about 3 inches (8 cm.) and the female 10 inches (26 cm.).

The *Trichocephalus crenatus* or *whip worm* is slender or hair-like in its anterior two-thirds and thick posteriorly. It is from 1.5 to 2 inches (40 to 45 mm.) long. It is found in the caecum attached to the wall by the hair-like portion.

The *OEsophagostomum dentatum* or *pin worm* is from 0.3 to 0.6 inch (8 to 15 mm.) long. It is found in the large intestine .

The *symptoms* of intestinal worms are not very evident in the average drove of hogs. None of the other farm animals are such common hosts for intestinal worms as hogs. But it is only in extreme cases of infection by intestinal worms, and in stunted and poorly-cared-for hogs, that very noticeable symptoms of disease are manifested. We must not take from the above statement that it is unnecessary to resort to treatment unless in exceptional cases. Intestinal worms interfere with the growth of young hogs, and may irritate and inflame the intestine, causing chronic indigestion, nervous symptoms, and in some cases death. This irritated and inflamed condition of the intestine is best noted in the abattoir by the ease with which the wall of the intestine that contains large numbers of worms tears when handling it.

The treatment of intestinal worms in hogs is both preventive and medicinal. If the conditions in the pens and houses are such as to enable the eggs and embryos to live for a long time, or the surroundings are favorable for infection of the animals through their feed and water supply, the herd may become badly infested with intestinal parasites. The preventive treatment consists in keeping hogs in clean, well-drained yards or pastures, and feeding them from clean troughs and concrete feeding floors that can be washed, when necessary, in order to keep them clean. Turpentine, given in a milk emulsion, is a common remedy for intestinal worms in hogs. The dose is one teaspoonful for every eighty pounds weight. This dose should be repeated daily for three days. The following vermifuge can be recommended: Santonin three to five grains, calomel five to eight grains. This is sufficient for one hundred pounds weight. If the pigs are small and it requires two or three to weigh one hundred pounds, the large dose should be given. If the hogs weigh one hundred pounds or more, they should receive the small dose. The drove should be divided into lots of ten or fifteen hogs each. The drugs should be mixed and divided into the same number of powders as there are lots of hogs. Ground feed is placed in the trough, dampened with milk, or water and the powder sprinkled evenly over it. The hogs are then allowed to eat the feed. It is best to dose them in the morning after they have been off feed for ten or twelve hours.

VERMINOUS BRONCHITIS IN CALVES.—The lung worms of cattle, *Strongylus micrurus* and *Strongylus pulmonaris*, may cause heavy losses in calves and yearlings. Older cattle may harbor these parasites, but they do not seem to be inconvenienced by them. The *S. micrurus* is from 1 to 3 inches (25 to 75 mm.) long. The *S. pulmonaris* is smaller. It is from 0.4 to 1.3 inches (10 to 35 mm.) long. They are found in the trachea and small bronchial tubes, where they are mixed with mucous secretions from the inflamed lining membrane of the bronchial tubes.

Wet seasons and low, wet pastures are said to favor the development of lung worms. Their life history is not fully understood. They do not persist generation after generation in the air passages of an animal, but the eggs and embryos are expelled and live for a time outside of the animal, when they may again become parasites of another or the same host.

The symptoms are the same as occur in bronchitis and pneumonia. Calves and yearlings are the only animals in the herd that may show symptoms of the disease. The air passages become irritated and inflamed, and the calf shows a slight cough. As the inflammation increases and the worms and mucous secretions plug up the small bronchial tubes, the coughing spells become more severe and rattling, wheezing sounds may be heard on auscultating the lungs. The calf finally loses its appetite, becomes emaciated and weak, and wanders off alone. It is usually found lying down and shows labored breathing that is occasionally interrupted by paroxysmal coughing. The death rate in poorly-cared-for herds is heavy.

VERMINOUS BRONCHITIS AND PNEUMONIA OF SHEEP.— The two lung worms of sheep are the *Strongylus filaria* and Strongylus_ rufescens_. The former is from 1.3 to 3 inches (33 to 80 mm.) long, and the latter from 0.6 to 1 inch (16 to 25 mm.) long. The *S. filaria* is thread-like and the *S. refuscens* hair-like in appearance. For this reason they are termed thread and hair lung-worms. The thread-worm is found in the trachea and the larger bronchial tubes, and the hair-worm in the most minute as well as the larger bronchioli.

This disease is most common in wet seasons. Undrained pastures and ponds are said to favor the spread of the disease. Permanent pastures favor the reinfection of the flock from year to year. The eggs and embryos are expelled in coughing, and live for a time in the pastures, pens and houses. The sheep become infected through the dust, drinking water or feed.

The symptoms of verminous bronchitis and pneumonia are quite characteristic. Lambs suffer most from these diseases. A number of animals in the flock are affected. Coughing, rapid and labored breathing, loss of appetite, emaciation and weakness are the usual symptoms noticed. When a paroxysm of coughing occurs, considerable mucus is expelled. An examination of the expectorations may result in finding a few lung worms. In poorly-cared-for flocks, and when complicated by stomach and intestinal worms, the death rate is usually heavy.

The treatment of lung-worm diseases in lambs and calves is largely preventive. We should use every possible precaution against introducing the infection into the herd or flock. It is not advisable to bring animals from an infected herd onto the premises, without subjecting them to a careful examination and a long quarantine before allowing them to stable or pasture with the other animals. Calves or lambs that show marked symptoms of disease should be given comfortable quarters, and special care and feeding. The entire herd or flock must be given the best care and ration possible. This is the only satisfactory method of treatment. Changing the pasture or lot frequently may help in ridding the premises of the infection.

VERMINOUS BRONCHITIS IN HOGS.—The lung worm, *Strongylus paradoxus,* is a common parasite of young hogs. It is from 0.6 to 1.6 inches

(16 to 40 mm.) long. When the infection is light, the worms are found mostly in the bronchial tubes of the margin and apex of the lung.

Infection with this parasite does not depend on the humidity of the soil, or low, wet pastures containing ponds. Probably dusty quarters are responsible in large degree for this disease.

The symptoms are most evident in pigs weighing from forty to eighty pounds. The first symptom is a cough, occurring on leaving the bed, after exercise and after eating. In badly infected cases the paroxysm of coughing is quite severe. The appetite usually remains good and the thriftiness of the pig is not seriously interfered with. The feeding of a suitable ration, and the good care that is usually given young hogs, are responsible for the mildness of the disease.

The treatment that is of most importance is clean quarters, and the feeding of a ration that will keep the pig growing and healthy. The sleeping quarters should be kept free from dust. Disinfectants should be used freely about the quarters.

THE KIDNEY WORM OF HOGS.—*Sclerostoma pinguicola* is the kidney worm of hogs. It is from 1 to 1.5 inches (25 to 27 mm.) long, and when seen against the kidney fat it appears dark or mottled. It is usually found in the fat in the region of the pelvis of the kidney. Although the kidney worm is capable of causing inflammatory changes in the tissues surrounding the kidney and the pelvis of this organ, the disease cannot be determined by any noticeable symptom. Paralysis of the posterior portion of the body is attributed to the presence of kidney worms by stockmen. There are no data by which we may prove that the kidney worm is responsible for this disorder.

The treatment is preventive. Clean feed, pens, watering troughs and feeding floors are the preventive measures indicated here. It is useless to attempt treatment with drugs, as the worms are out of reach of any drug that may be administered.

WORMS OF THE DIGESTIVE TRACT OF POULTRY.—Poultry are often seriously infested with worms. A small number of the less injurious worms may not cause any appreciable symptoms of disease; but the fowl that harbors them is a source of infection to the other fowls. The infectious nature of parasitic disease caused by worms should be recognized more fully than at present by poultrymen.

The different species of poultry are hosts for many different species of round-worms, thorn-headed worms and tapeworms. Dr. Kaupp states that *Acaris inflexa* or large round-worm, *Heterakis pipilosa* or small round-worm, and the *Spiroptera hamulosa* or gizzard-worm are frequently found in fowls. The common round-worm may be found in the first portion of the intestine, and the small round-worm in the caecum. Neither of the species are dangerous unless present in large numbers. They may then obstruct the intestine, and irritate the intestinal mucous membrane. This may cause constipation, catarrhal inflammation of the intestine and diarrhoea. The gizzard-worm is the most dangerous of the parasites mentioned. The gizzard has an important digestive function, and any condition that may weaken its muscular walls may cause serious digestive disorders. This parasite may encyst in the wall of the gizzard.

The treatment of intestinal worms in poultry is both preventive and curative. The preventive measures consist in keeping the houses and runs clean. Air-slaked lime should be scattered over the runs every few weeks. The drinking places should be cleaned and disinfected daily. All possible precautions should be taken in order to prevent filth from getting into the drinking water. Epsom salts, powdered areca nut and santonin are the remedies commonly recommended for the treatment of intestinal worms. From twenty to forty grains of Epsom salts may be given. Powdered areca nut is recommended in from three to ten grain doses. Santonin may be given in from one to two grain doses. Both the areca nut and santonin may be given with the feed.

THE GAPES IN BIRDS.—The gape-worm, *Syngamus trachealis*, is from 0.2 to 0.8 inch (5 to 20 mm.) long. The male and female are permanently united. The male is about one-third as long as the female, and when attached to the anterior third of the female, gives the pair a forked appearance.

Fowls become infested with the gape-worm by eating the adult parasite that has been expectorated, or an earth worm that is host for the immature parasite. The embryo gape-worm is freed in the intestine, and from here they are supposed to migrate into the abdominal air sacs and to the trachea and bronchi.

The symptom are most severe in very young fowls. The affected bird opens its mouth and appears to gasp for breath, sneeze and attempt to swallow. In the severe cases the appetite is interfered with, mucus accumulates in the mouth and the bird is dull and listless. The death rate is quite high in young-chickens and turkeys.

The treatment is both preventive and curative. If the gape-worm is known to be present in the runs, the ground should be covered with lime, and the fowls moved to fresh runs if possible. The young birds should not be exposed to the infection until they are well feathered out. Antiseptics may be given with the drinking water. Disinfectants should be used freely about the poultry houses, and the quarters kept clean. The worms may be snared by inserting a stiff horse hair that has been twisted and forms a loop into the trachea. This may be dipped into camphorated oil or turpentine. This treatment should be repeated until the bird has been relieved.

QUESTIONS

1. Name the different groups of internal parasites; give examples of each.

2. What conditions favor liver rot? Give the life history of the liver fluke.

3. Name three common tapeworms; give the life history of the beef and pork tapeworms.

4. Name the common intestinal worms of horses and give the treatment.

5. Give the symptoms and treatment of stomach-worm disease of sheep.

6. Name the common intestinal worms of hogs and give treatment.

7. What species of domestic animals suffer most of verminous bronchitis? Give the treatment.

8. Name the common internal parasites of poultry and give treatment.

PART VI
INFECTIOUS DISEASES

CHAPTER XXVI
HOG-CHOLERA

HOG-CHOLERA is a highly infectious disease of swine. It is characterized by an inflammation, of the lymphatic glands, kidneys, intestines, lungs and skin. The inflammation is hemorrhagic in character, the inflamed organs usually showing deep red spots or blotches.

Hog-cholera is especially prevalent in the corn-raising States which possess a denser hog population than any other section of the United States. In this country the loss from hog-cholera in 1913 amounted to more than $60,000,000, and it may be considered of greater economic importance than any of the other animal diseases.

SPECIFIC CAUSE.—The specific cause of hog-cholera is an *ultra-visible organism* that is present in the excretions, secretions and tissues of a cholera hog. De Schweinitz and Dorset in 1903 produced typical hog-cholera by inoculating hogs with cholera-blood filtrates that were free from any organism that could be demonstrated by microscopical examination or any cultural method. The term ultra-visible virus is applied to the virus of hog-cholera.

The ultra-visible virus is eliminated from the body of the cholera hog with the body secretions and excretions. Healthy hogs contract the disease by eating feed or drinking water that is infected with the virus. There are other methods of infection, but field and experimental data show that hog-cholera is commonly produced by taking the germs into the body with food and drinking water.

ACCESSORY CAUSES.—The usual method of introducing hog-cholera into a neighborhood is through the importation of feeding or breeding hogs that were infected with the disease before they were purchased, or became infected through exposure to the disease in the public stock-yards and stock-cars. The shipping of feeding hogs from one section of the country to another, and from public stock-yards, has always been productive of hog-cholera. Dr. Dorset states that more than fifty-seven per cent of the hog-cholera outbreaks are caused by visiting, exchanging work, exposure on adjoining farms and harboring the infection from year to year , and more than twenty-three per cent to purchasing hogs and shipping in infected cars, birds and contaminated streams.

In neighborhoods where outbreaks of hog-cholera occur necessary precautions against the spread of the disease are not taken. The *exchange of help* at threshing and shredding time in neighborhoods where there is an outbreak of hog-cholera is the most common method of spreading the disease. *Visiting farms* where hogs are dying of cholera; walking or driving a team and wagon through the cholera-infected yards; stock buyers, stock-food and cholera-remedy venders that visit the different farms in a neighborhood may distribute the hog-cholera virus through the infected filth that may adhere to the shoes, horses' feet and wagon wheels. *Cholera hogs* may carry the disease directly to a healthy herd when allowed to run at large. *Streams* that are polluted with the drainage from cholera-infected yards are common sources of disease.

Pigeons, dogs, cows and *buzzards* that travel about the neighborhood and feed in hog yards and on the carcasses of cholera hogs may distribute the disease. Because of the active part that dogs, birds and surface drainage take in the distribution of hog-cholera, the practice of allowing the carcasses of dead hogs to lie on the ground and decompose is responsible for a large percentage of the hog-cholera outbreaks.

Age is an important predisposing factor. Young hogs are most susceptible to cholera, and this susceptibility can be greatly increased by giving them crowded, filthy quarters. Infection with lice, lung and intestinal worms, the feeding of an improper ration and sudden changes in the ration lower the natural resistance of a hog against disease. Pampered hogs usually develop acute cholera when exposed to this disease.

Hog-cholera is more virulent or acute during the summer and fall months than it is during the winter and spring months. After the disease sweeps over a section of country, it becomes less virulent and takes on a subacute or chronic form. Outbreaks of hog-cholera usually last two or three years in a neighborhood. This depends largely on the number of susceptible hogs that were not exposed to the infection the first season, and the preventive precautions observed by the owners.

PERIOD OF INCUBATION.—The length of time elapsing between the exposure of the hog to the cholera virus, and the development of noticeable symptoms of hog-cholera, varies from a few days to two or three weeks. The length of this incubation period depends on the susceptibility of the animal, the virulence of the virus and the method of exposure. An acute form of hog-cholera indicates a short period of incubation, and a chronic form, a long period.

SYMPTOMS.—The symptoms of hog-cholera may differ widely in the different outbreaks of the disease. The symptoms may be classified under the following forms: Acute, subacute and chronic. The acute form of hog-cholera is the most common. The early symptoms are tremors, fever, depressed appearance, marked weakness, staggering gait, constipation and diarrhoea, labored breathing and convulsions. Death may occur within a few hours or a few days. Recovery seldom occurs. In the subacute form, the symptoms are mild and develop slowly. Recovery may take place within a few days, or after extending over a week or ten days it may assume the chronic form. Very often in outbreaks of subacute cholera a large majority of the herd does not show visible symptoms of the disease. In the chronic form, marked symptoms of pleuropneumonia and chronic inflammation of the intestine are common. Ulcers and sores form on the skin and the hair may come off. Large portions of the skin may become gangrenous and slough. Young hogs are usually stunted and emaciated.

The first symptom of disease is an elevation of body temperature.

At the beginning of any outbreak of hog-cholera the *body temperatures* of the apparently healthy animals may vary from 105° to 108° F. After a few days, animals that are fatally sick or recovering from the disease may show normal or subnormal body temperature.

Loss of appetite is the first symptom of disease usually noted by the person in charge of the herd. The hog may show a disposition to eat dirt. The sick hog is usually found lying in its bed, or off by itself in a quiet place. It presents a rather *characteristic appearance*. The back is arched, the hind feet are held close together, or crossed, the abdomen is tucked up and the hog appears weak in its hind parts. *Diarrhoea* or *constipation* may be present. The color of the diarrhoeal discharges varies according to the character of the feed, and it may be more or less tinged with blood and have a disagreeable odor. The urine is highly colored.

The respirations and pulse beats are quickened and abnormal in character. Thumps sometimes occur. When the mucous membranes lining the throat and anterior air passages are thickened, the respirations are noisy and difficult. The animal may cough on getting up from its bed and moving about. There is at times a noticeable discharge from the nostrils. When the *lungs* are inflamed the respirations are quickened and labored. In case the pleural membrane is inflamed, the respiratory symptoms are more severe, and the hog shows evidence of pain when the walls of the chest are pressed on. The *pericardium* may be inflamed. In such cases the hog staggers and falls when forced to walk.

The central nervous system may be involved by the inflammation. The usual symptoms occurring in inflammation of the brain and its coverings are then present. A sleepy, comatose condition may end in death, or the animal dies in a convulsion.

The secretions of the skin and mucous membranes are abnormal. The skin in the regions of the ears, inside of the thighs and under surface of the body is moist, dirty or discolored red. Just before death the skin over the under surface of the body becomes a purplish red. In the chronic form, a dirty, thickened, wrinkled skin is commonly observed. At first the secretion from the eyes is thin and watery, but it becomes thick, heavy and pus-like, causing the margins of the lids to adhere to each other.

The death rate in hog-cholera varies in the different forms of the disease. The average death rate is about fifty per cent.

DIFFERENTIAL DIAGNOSIS.—The diagnosis of hog-cholera in the field must depend on the clinical symptoms, post-mortem lesions and history of the outbreak. The history should be that of a highly infectious disease.

Abnormal body temperatures of a large percentage of the herd indicate the presence of an acute infectious disease. We should then destroy one of the sick hogs and make a careful post-mortem examination . An early diagnosis of the disease is necessary, as this enables us to use curative treatment when it will do some good, and take the necessary steps toward preventing the spread of the disease to neighboring herds.

Intestinal and lung worms are common in young hogs. The presence of these worms does not always indicate that they are the cause of the sickness and death of the animal. Such parasites are injurious and may cause disease, but it is only in rare cases that they cause death.

"Pig typhoid" is sometimes spoken of as a highly infectious disease involving the intestines. A disease of hogs that may be termed typhus-fever sometimes affects a large number of the hogs in the herd. This disease occurs among hogs kept in small yards and houses that are crowded, unsanitary and in continuous use, or when the hogs drink from wallows, ponds and creeks.

The term swine-plague should not be used in speaking of outbreaks of hog-cholera, as it is now considered a form of hog-cholera involving especially the lungs.

LESIONS.—In *acute hog-cholera* the inflammation is hemorrhagic in character. Small, red spots and blotches occur in different organs and tissues. In the *chronic form* of the disease ulceration of the intestinal and gastric mucous membrane, inflammation of the lungs and pleura and sloughing of the skin are common lesions.

The skin over the under side of the neck, body and inside of the thighs may appear red or purplish-red in color. The different groups of *lymphatic glands* are enlarged and softened. They may vary in color from a grayish-red to a deep red, depending on the degree of engorgement with blood. The pleura and pericardium may show small red spots and blotches. The *kidneys* are usually lighter colored than normal, and marked with red spots and blotches . The *spleen* may show no evidence of disease. It may be large and soft, or even smaller than normal. The *liver* may be enlarged and dark, or mottled and light colored.

The *stomach* and *intestines* may show hemorrhagic spots and blotches. Sometimes the gastric and intestinal mucous membrane is a brick red. Ulceration of the mucous membrane is common .

Small, red spots may be present on the surface of the *lungs* . Scattered lung lobules or a large portion of the lungs may be inflamed. In chronic hog-cholera, pleural exudation, adhesions and abscesses in the lung tissue may occur. Inflammations of the pericardium and heart muscle are less common lesions.

PREVENTIVE MEASURES.—Hog-cholera is the most widespread infectious disease of hogs, and all possible precautions against its distribution to healthy herds should be practised. Hogs coming from other herds and stock shows should be excluded from the home herd until they are positively shown to be free from disease. They should be quarantined in yards set off for this purpose. The hogs should be cleaned by dipping or washing them with a disinfectant. The quarantine period should be longer than the average period of incubation. Three weeks is sufficient.

The possible introduction of the disease into the pens by people, dogs, birds and other carriers of the disease should be guarded against, especially if cholera is present in the neighborhood. The exchange of help at threshing and shredding time with a neighbor who has hog-cholera on his farm is a common method of distributing the infection. It is not advisable to allow a stranger to enter your hog-houses and yards, unless his shoes are first disinfected. Whenever it is necessary for a person to enter yards where the disease is present, the shoes should be cleaned and disinfected on leaving. The wheels of wagons, and the feet of horses that are driven through cholera yards, should be washed with a disinfectant. The feet of feeding cattle that are shipped from stock-yards should be treated in the same manner. Persons taking care of cholera hogs should observe the necessary precautions against the distribution of the disease, and see that others practise like precautions.

Hog-yards should be well drained and all wallow holes filled. Pens and pastures through which the drainage from the swine enclosures higher up flows should not be used for hogs.

CARE OF A DISEASED HERD.—When an outbreak of hog-cholera occurs on a farm the farm should be quarantined. The herd should be moved away from running streams, public roads and line fences, so that neighboring herds are not unnecessarily exposed to the disease. During the hot weather shade and an opportunity to range over a grass lot or pasture are highly necessary. A recently mowed meadow, or a blue grass pasture and a low shed, open on all sides and amply large for the herd to lie under, give the animals clean range and comfortable, cool quarters. Roomy, dry, well-ventilated sleeping-quarters that are free from drafts and can be cleaned and disinfected are best when the weather is cold and wet.

In the subacute, and in the early part of an acute outbreak of hog-cholera, it is advisable to separate the sick from the well hogs. The fatally sick animals should be destroyed.

A very light ration should be fed and an intestinal antiseptic given with the feed. A thin slop of shorts is usually preferred. Four ounces of pulverized copper sulfate may be dissolved in one gallon of hot water, and one quart of this solution may be added to every ten gallons of drinking water and slop. Water and slop should not be left in the troughs for the hogs to wallow in. The troughs should be disinfected and turned bottom side up as soon as the hogs have finished feeding and drinking. Kitchen slop and sour milk should not be fed. The care and treatment of the herd require work and close attention on the part of the attendant. Indifferent, careless treatment is of no use in this disease.

A disinfectant should be sprayed or sprinkled about the feed troughs, floors, pens and sleeping quarters daily.

DISPOSING OF DEAD HOGS.—The carcasses of the dead hogs should be burned. Before placing the carcass on the fire, it should be cut open and several long incisions made through the skin. A crematory may be made by digging two cross trenches that are about one foot deep at the point where they cross, and shallow at the ends. Iron bars or pipe may be laid over the trenches where they cross for the carcass to rest upon, or woven wire fencing securely fastened with stakes may be used in the place of the iron bars. If the carcass is disposed of by burying, it should be buried at least four feet deep and covered with quicklime.

DISINFECTING THE YARDS AND HOUSES.—If the sick hogs are moved to new quarters at the beginning of the outbreak, the hog houses and yards should be cleaned and disinfected . The manure and all other litter should be hauled away to a field where there is no danger from this infectious material becoming scattered about the premises, leaving a centre of infection in the neighborhood and causing outbreaks of cholera among neighboring herds. It may be advisable to burn the corn-cobs and other litter that have accumulated about the yards. Loose board floors should be torn up and the manure from beneath removed. Portable houses should be removed. The floors, walls of the house and fences should be first cleaned by scraping off the filth, and then sprayed with a three per cent water solution of a cresol or coal tar disinfectant to which sufficient lime has been added to make a thin whitewash. Three or four months of warm, sunny weather are sufficient to destroy the cholera infection in well-cleaned yards.

ANTI-HOG-CHOLERA SERUM.—The credit of developing the first and at present the only reliable anti-hog-cholera serum and method of vaccination belongs to Drs. Dorset and Niles. Anti-hog-cholera serum came into general use in 1908, and all of the swine-producing States have established State laboratories for the production of this serum.

Anti-hog-cholera serum is produced by injecting directly, or indirectly, into the blood-vessels of an immune hog a large quantity of cholera virus, secured by bleeding a hog that is fatally sick with acute cholera, and bleeding the injected animal after it has completely recovered from the injection. The injection of the cholera blood is for the purpose of stimulating the production of antibodies by the body tissues, and raising the protective properties of the immune hog's blood. An animal so treated is called a hyperimmune . The blood from the hyperimmunes is defibrinated and a preservative added, and after it has been tested for potency and freedom from contaminating organisms, it is ready for use.

THE VACCINATION OF HOGS WITH ANTI-HOG-CHOLERA SERUM. — The vaccination of a hog by the single method consists in injecting hypodermically or intramuscularly anti-hog-cholera serum. The immunity conferred may not last longer than three or four weeks.

The vaccination of a hog by the *double method* consists in injecting hypodermically or intramuscularly anti-hog-cholera serum and hog-cholera blood.

The vaccination or treatment of a cholera hog showing noticeable symptoms, or a high body temperature, consists in injecting hypodermically or intramuscularly anti-hog-cholera serum .

The region into which the serum and cholera blood may be injected are the inside of the thigh, within the arm, flank and side of the neck . Two hypodermic syringes, holding about twenty cubic centimetres and six cubic centimetres, and having short, heavy, seventeen or eighteen-gauge slip-on needles, should be used. The small syringe is used for injecting the virulent or cholera blood which is injected into a different part than the serum. The quantity of serum and virus injected varies with the size and condition of the animal. *Young hogs* should receive one-half cubic centimetre of serum for each pound of body weight, and *cholera hogs* should be given one-half more to twice the dose that is recommended for healthy animals. The dose of virus recommended varies from one to two cubic centimetres for each hog.

In vaccinating *small pigs* not more than five, and in large hogs not more than twenty, cubic centimetres should be injected at any one point. The *body temperature* of each animal should be taken. A body temperature of 103.5\260 F. in a mature hog and a body temperature of 104\260 F. in a young hog may indicate hog-cholera. Exercise, feeding and close confinement in a warm place may raise the body temperature above the normal.

Hogs that are to be vaccinated or treated should not be given feed for at least twelve hours before handling them. If possible they should be confined in a roomy, clean, well-bedded pen. If this is practised, they are cleaner and easier to handle and their body temperatures are less apt to vary. After the treatment or vaccination the hogs should be fed a light diet for a period of at least ten days, and the ration increased gradually in order to avoid causing acute indigestion. This is necessary because of the elevation in body temperature resulting from the inability of the animal to digest heavy feeds, kitchen slops and sour milk. If poor judgment is used in caring for the vaccinated hogs, and the person who vaccinates them uses careless methods, heavy losses from acute indigestion, blood poisoning, or hog-cholera may occur.

QUESTIONS

1. What is the specific cause of hog-cholera? Give and describe the different methods of spreading the disease.

2. What are the symptoms of hog-cholera?

3. Give the preventive and curative treatment of hog-cholera.

4. What is anti-hog-cholera serum? Give the different methods of vaccination and treatment.

CHAPTER XXVII
TUBERCULOSIS

Tuberculosis is a contagious an and domestic animals, affecting any the lymphatic glands and lungs, change in the tissues is the formation tubercle or nodule.

HISTORY.—Tuberculosis is one of the oldest of known diseases of domestic animals and man. Its contagious character was proven by Villemin in 1865, who by experential infection transmitted tuberculosis from man to animals and from animal to animal. It was in 1882 that Dr. Robert Koch discovered and proved by inoculation experiments that the disease was caused by a specific germ . Prior to the experiments by Villemin and Koch, the belief was that tuberculosis was due to heredity, unsanitary conditions and inbreeding. Following discovery of the specific germ and conditions favoring its development and spread, numerous scientifically conducted experiments were made. These resulted in practical methods of control and elimination of tuberculosis from herds having this disease. By carefully conducted experiments and other forms of educational work the infectious character of tuberculosis and the economic importance of preventative measures have been demonstrated. The average stockman is well informed regarding the character and economic importance of this disease, but there is no general application of this knowledge, and tuberculosis is increasing in dairy and breeding herds. The slow development of tuberculosis, and the absence of visible symptoms during the early stage of the disease, are responsible for this condition and the extensive infection of dairy and breeding herds.

PREVALENCE OF THE DISEASE.—Tuberculosis is very prevalent among cattle and swine in all countries where intensive agriculture is practised. It is a rare disease among cattle of the steppes of eastern Europe and the cattle ranges of the western portion of the United States. In countries where dairying is an important industry, tuberculosis is a common disease of cattle and hogs. The abattoir reports of Europe and the United States show that tuberculosis is on the increase among domestic animals. The Bureau of Animal Industry of the United States Department of Agriculture reports that out of 400,008 cattle tested with tuberculin 9.25 per cent reacted.

Melvin states that the annual loss from tuberculosis in the United States is about $23,000,000. In dairy herds in which the disease has existed for several years, it is not uncommon to find from 25 to 75 per cent tubercular.

THE DIRECT CAUSE.—The direct cause of tuberculosis is Koch's *Bacillus tuberculosis*. This is a slender, rod-shaped microorganisms occurring in the diseased tissues, feces and milk of a tubercular animal. It belongs to that small group known as acid-fast bacteria. The tubercle bacillus is not really destroyed by external influences, and it may retain its virulence for several months in dried sputum if protected from the light. Its vitality enables it to resist high temperatures, changes in temperature, drying and putrefaction to a, greater degree than most non-spore-producing germs. Direct sunlight destroys the germ within a few hours, but it may live in poorly lighted, filthy stables for months. A temperature of 65\260 C. destroys it in a few minutes.

Animals that, have advanced or open tuberculosis may disseminate the germ of the disease in the discharge from the mouth, nostrils, genital organs, in the intestinal excreta and milk. The germs discharged from the mouth and nostrils are coughed up from the lungs and may infect the feed. Milk is a common source of infection for calves and hogs. Allowing hogs to run after cattle is sure to result in infection of a large percentage of them, if there are any open cases of tuberculosis in the herd.

PREDISPOSING CAUSES.—Any condition that may lessen the resistance of the body or enable the tubercle bacillus to survive the exposure outside the body favors the development of the disease and the infection of the healthy animals. Crowded, poorly ventilated, filthy stables lower the disease-resisting power of the animal, and favor the entrance of the germs into the body. Under such unsanitary conditions, tuberculosis spreads quickly among dairy cattle, and a large percentage of the animals develop the generalized form of the disease. Sanitary stables and yards do not prevent the spread of the disease among animals that live in close contact with one another. Fresh air and sanitary surroundings only check the spread and retard its progress.

INTRODUCTION OF TUBERCULOSIS INTO THE HERD.—The common method of introducing tuberculosis into the herd is through the purchase of animals having the disease. Such animals may be in apparent good health at the time of purchase, and be affected with generalized or open tuberculosis.

A source of infection is by unknowingly buying cows that have reacted to the tuberculin test. The indiscriminate use and sale of tuberculin are largely responsible for the large number of reacting animals that have been placed on the open market. This dishonest practice has resulted in the rapid spread of the disease in certain localities. For years a large percentage of the breeding herds have been infected, and the writer has met with several herds of dairy and beef cattle that became tubercular through the purchase of tubercular breeding animals.

SYMPTOMS.—There is no one symptom by which we may recognize tuberculosis in cattle and hogs. None of the symptoms shown by a tubercular animal are characteristic, unless it is in the late stage of the disease. In a well-cared-for animal, the lymphatic glands in the different regions of the body, the lungs, liver and other organs, may be full of tubercles without causing noticeable symptoms of disease .

Tuberculosis may attack any organ of the body, and in the different cases of the disease the symptoms may vary. Enlargement of the glands in the region of the throat, and noisy, difficult breathing are sometimes present. The udder frequently shows hard lumps scattered through the gland. Bloating may occur if a diseased gland in the chest cavity presses on the oesophagus and prevents the usual passage of gas from the paunch. Chronic diarrhoea may occur. If the disease involves the digestive tract, the animal is unthrifty and loses flesh rapidly. Coughing is not a characteristic symptom, and we should not place too much emphasis on it. If the lungs become tubercular the animal usually has a slight, harsh cough. The cough is first noticed when the cattle get up after lying down, when the stable is first opened in the morning and when the animals are driven. If the chest walls are thin, soreness from pressure on the ribs may be noted. By applying the ear to the chest wall and listening to the lung sounds, absence of respiratory murmurs and abnormal sounds may be distinguished, due to consolidation of the lung tissue, abscess cavities and pleural adhesions. In a well-advanced case the hair is rough, the skin becomes tight and the neck thin and lean. The animal may breathe through the mouth when it is exercised. Weakness may be a prominent symptom.

Breeding animals that are well fed and cared for may live for several years before showing noticeable symptoms of tuberculosis. The disease progresses more rapidly in milch cows, especially if given poor care. Calves allowed to nurse a tubercular mother that is giving off tubercle bacilli frequently develop enlarged throat glands and the intestinal form of the disease.

Hogs develop a generalized form of tuberculosis more quickly than cattle, but an unthrifty, emaciated condition is seldom noted in hogs under ten months old.

POST-MORTEM LESIONS.—The effect of the tubercle bacillus on the body is to irritate and destroy the tissues. Lumps or tubercles form in the lymphatic glands, liver, lungs, spleen , serous membranes, kidneys and other body organs (Figs. 91 and 92). The tubercles may be very small at first, but as the disease progresses they continue to enlarge until finally a tubercular mass the size of a base-ball, or larger, is formed (Figs. 93, 94, 95 and 96). Lymphatic glands may become several times larger than normal and the liver and lungs greatly enlarged. The pleura and peritoneum may be thickened and covered with tubercles about the size of a millet seed, or larger. Pleural and peritoneal adhesions to the organs within the body cavities are common.

The tubercle usually undergoes a cheesy degeneration. Old tubercles may become hard and calcareous. Sometimes the capsule of the tubercle is filled with pus. A yellowish, cheesy material within the capsule of the tubercular nodule or mass is typical of the disease.

THE TUBERCULIN TEST.—The only certain method of recognizing tuberculosis is by this test. There is no other method of recognizing this disease that is more accurate than the above test.

The substance used in testing animals for tuberculosis is a laboratory product. It is a germ-free fluid prepared by growing the tubercle bacillus in culture medium (bouillon) until charged with the toxic products of their growth. The culture medium is then heated to a boiling temperature in order to destroy the germs. It is then passed through a porcelain filter that removes the dead germs. The remaining fluid is tuberculin.

There are two methods of applying the tuberculin test. The subcutaneous test consists in injecting a certain quantity of tuberculin beneath the skin, and keeping a record of the body temperature of the animal between the eighth and eighteenth hours following the injection. Tubercular animals show an elevation in temperature that comes on about the eighth or twelfth hour of the test. In the *intradermal test,* a small quantity of a special tuberculin is injected into the deeper layer of the skin. The seat of the injection in cattle is a fold of the skin on the under side of the base of the tail. In tubercular animals the injection is followed by a characteristic local swelling.

The control of tuberculosis is largely in the hands of the breeder and dairyman. This is a disease that requires the cooperation of stockmen and sanitary officers in the application of control measures. If there are several open cases of tuberculosis in a herd of cattle, the application of the tuberculin test, removal of the reacting animals and disinfection of the premises are not sufficient to eradicate the disease. It is necessary to repeat the tuberculin test within six months, and later at twelve-months intervals, until none of the animals that remain in the herd react.

The most practical method of disposing of dairy cows that react to the tuberculin test is to slaughter them. Unless a large percentage of the herd is tubercular, it is not advisable to practise segregation and quarantine. This may be advisable if the reactor is a valuable breeding animal, unless visible symptoms are shown. The milk from reacting cows may be used if it is boiled or sterilized. Whenever a calf is born of a reactor, it should be separated from the mother and fed milk from a healthy cow.

The separation of the tubercular from the healthy cows must be complete. Separate buildings, yards and pastures that do not join the quarters where the healthy animals are kept should be provided. The person attending the reactors should not attend the healthy animals, and separate forks, shovels, pails and other utensils should be provided for the two herds.

The best method of controlling tuberculosis in hogs is to slaughter all reactors, disinfect yards and houses and move the herd. If the old quarters are free from filth and carefully disinfected, the hogs may be returned without danger of infection after six months. A retest of the herd should be made before returning them to the permanent quarters and the reactors slaughtered.

QUESTIONS

1. Give the history of the early experimental work in tuberculosis; give the common methods of spreading the disease.

2. What are the symptoms and post-mortem lesions in tuberculosis?

3. Give the method of controlling tuberculosis.

CHAPTER XXVIII
INFECTIOUS DISEASES COMMON TO THE DIFFERENT SPECIES OF DOMESTIC ANIMALS

SEPTICAEMIA AND PYAEMIA.—The term commonly used in speaking of simple septicaemia and pyaemia is blood poisoning.

These infectious diseases are *caused* by several different species of bacteria that gain entrance to the tissues by way of wounds. The bacteria that cause pyaemia are transferred by the blood stream to different organs and produce multiple abscesses. In septicaemia, the bacteria may occur in immense numbers in the blood and produce a general infection of the tissues, causing a parboiled appearance of the liver, heart, voluntary muscles and kidneys, and enlargement of the spleen. The two forms of infection are often present at the same time.

The forms of bacteria that may cause blood poisoning are the *Staphylococcus pyogenes aureus* and *albus* , *Streptococcus pyogenes* , *Bacillus pyocyaneus*, *Bacillus coli communis*, and the bacillus of malignant oedema (Figs. 99 and 100). The latter is included with the bacteria that produce blood poisoning because it is a frequent cause of wound septicaemia. Subcutaneous, punctured, lacerated, contused and deep wounds without suitable drainage are the most suitable for the development of and infection of the tissues with the above germs. Wound infection is most common during hot weather.

The symptoms are both general and local. The tissues in the region of the wound become swollen and painful. In malignant oedema the swelling pits on pressure, and if the wound is open, the surface becomes soft and may slough. The body temperature may be several degrees above the normal, the appetite is impaired or the animal stops eating and acts sleepy. A small amount of highly-colored urine may be passed. Nervous symptoms, such as muscular twitching, excited condition, delirium and paralysis, may be noted.

The prognosis is unfavorable. In pyaemia the animal may live from a few days to several months. Septicaemia usually terminates fatally in from two to ten days.

The treatment is largely preventive. Wounds should be given prompt attention. They should be freed from all foreign substances and washed with a disinfecting solution. A contused-lacerated wound should not be sutured if this interferes with the cleansing of it, and the escape of the wound secretions. All punctured wounds should be enlarged so as to permit of treatment and drainage.

HEMORRHAGIC SEPTICAEMIA.—An acute infectious disease of ruminants and swine, characterized by hemorrhages in the different body tissues that appear as small red spots or blotches.

The specific cause of this disease is the *Bacillus bovisepticus* . This bacillus probably enters the body tissues by way of the lining membrane of the intestinal and respiratory tracts. In the northern States, cattle pasturing on marsh lands and swampy pastures are more often affected with the disease in the late summer and fall than at other seasons of the year.

The drinking of contaminated surface water that collects in muddy pools and ponds may cause the disease. Cattle pasturing in stalk fields sometimes become infected in this way. Dusty sleeping quarters and small, crowded, muddy yards seem to favor the development of the disease in hogs. Exposure, insufficient exercise and careless feeding are the predisposing factors.

The symptoms vary according to the animal and organ, or organs of the body affected and the violence of the attack. The disease may be acute or subacute. The brain and its membranes, lungs and air-passages and intestines may become affected. The symptoms may be classed under the head of nervous, respiratory and intestinal and they may be very unsatisfactory from the standpoint of diagnosis. The history and post-mortem lesions are of most value in the recognition of this disease. The local conditions, the loss of several animals in the herd and the finding of hemorrhagic lesions in the different body tissues may enable the examiner to correctly diagnose the disease. It is very advisable in order to confirm the diagnosis to make a bacteriological examination of the tissues.

The acute form of the disease is very fatal. Animals that have the subacute form usually recover. The death-rate is between five and fifteen per cent of the herd. The mortality is heavier than this unless prompt preventive measures are taken.

Preventive treatment is of the greatest importance. Cattle that become affected when running on pasture should be moved, or in case a part of the pasture is swampy, we may prevent further loss by fencing off this portion. Drinking places that are convenient and free from filth should be provided. Watering troughs and drinking fountains should be cleaned and disinfected every few weeks. For this purpose, use a three per cent water solution of a cresol disinfectant, or a ten per cent water solution of sulfate of iron. Dusty quarters should be cleaned and disinfected. Dirt floors may be sprinkled with crude oil.

When an outbreak of septicaemia haemorrhagica occurs in a herd, both the well and sick animals should be given a physic. Cattle may be given one-half pound of Epsom salts, repeated in three or four days; sheep and hogs from one to four ounces of raw linseed oil. Animals that have the subacute form of the disease may be given stimulants, and iron and bitter tonics.

ANTHRAX, CHARBON.—This is an acute infectious disease affecting many different species of animals. Anthrax is one of the oldest animal diseases, and early in the history of the race it existed as a plague in Egypt. It most commonly affects cattle, sheep and horses. Man contracts the disease by handling wool and hides from animals that have died of anthrax, and by accidental inoculation in examining the carcass of animals that have died of the disease.

Cause.—Anthrax is caused by a rod-shaped, spore-producing microorganism, *Bacillus anthracis*. It gains entrance to the body by way of the intestinal tract, lungs and air-passages and the skin. The bites of insects play an important part in the distribution of the disease in some localities, but the most common method of infection is by way of the digestive tract, through eating and drinking food and water contaminated with the anthrax germs. The spores of the B. *anthracis* are very resistant to changes in temperature and drying. They may live for years in rich, moist inundated soils. River-bottom and swampy lands that have become infected with discharges from the bodies of animals sick with anthrax, and by burying the carcasses of animals that have died of this disease, retain the infection for many years. Anthrax is very widely distributed. It is the most prevalent in the southern portion of the United States, especially the lower portion of the Mississippi Valley.

The symptoms vary in different cases, depending on the organs affected, and the virulence and amount of virus introduced. The *apoplectic form* is very acute. The disease sets in suddenly; the animal trembles, staggers, falls and dies in a convulsion. Blood may be discharged from the nose and with the urine and faeces.

In the *abdominal form*, abdominal pain, diarrhoea, prolapse of the rectum, bloating and doughy swellings in the region of the abdomen occur.

In the *thoracic form*, the symptoms are bloody discharge from the nostrils, salivation, rapid, difficult breathing and swelling in the region of the throat. Local or skin lesions may occur in conjunction with, or independent of, the above forms of disease. These are carbuncles one or two inches in diameter that are hot and tender at first, but later become gangrenous, diffused swellings.

On post-mortem examination the blood is found tarry and dark, and bloody exudates may be found in the abdominal and thoracic cavities. The spleen is soft and two or three times larger than normal. The diagnosis should be confirmed by finding the *B. anthracis* in the blood and tissues. The death-rate is very high, usually about seventy-five per cent.

The treatment is preventive. Animals should be kept away from lots and pastures where deaths from anthrax have been known to occur, unless immunized against the disease. Marshy, swampy land that is infected with the germs of anthrax should be drained and cultivated.

When an outbreak of the disease occurs, all of the animals should be vaccinated. The carcasses of the animals that die should be cremated at or near the place where they die. If hauled or dragged, the necessary precautions should be taken against scattering the infectious material from the carcass, and plenty of disinfectants used. Persons attending the animals should be warned against opening or handling the carcass without protecting the hands with rubber gloves.

Anthrax vaccine should not be used by incompetent persons.

ULCERATIVE STOMATITIS. (ULCERATIVE SORE MOUTH.)—This is an infectious disease of young animals. Pigs from a few days to a few weeks of age are the most commonly affected.

The specific cause of ulcerative sore mouths is the *Bacillus necrophorus* . The infectious agent is distributed by the udder of the mother becoming soiled with filth from the stable floor and yards, and by affected pigs nursing mothers of healthy litters. Filth, sharp teeth and irritation to the gums from the eruption of the teeth are important predisposing factors.

The symptoms are, at first, an inflammation of the mucous membrane lining the lips and cheeks and covering the gums. The inflamed parts are first swollen and a deep red color; later, white patches form and the part sloughs, leaving a deep ulcer. As ulceration progresses, difficulty in nursing increases until finally the young animal is unable to suckle. If ulceration of the mouth is extensive, the animal may be feverish, dull and lose flesh rapidly. Portions of the lips, gums and snout may slough off. The death-rate in pigs is very high.

The preventive treatment consists in keeping the quarters and yards in a sanitary condition, and using all possible precautions against the introduction of the disease into the herd. The diseased young and mother should be separated from the herd and the quarters disinfected daily. The mouths of all the young should be examined daily and the diseased animals treated. The ulcers should be scraped or curetted and cauterized with lunar caustic, and the mouth washed daily with a two per cent water solution of a cresol disinfectant. Dipping pigs headforemost into a water solution of permanganate of potassium (one-half teaspoonful dissolved in a gallon of water), twice daily, may be practised if the herd is large.

It is usually most economical to kill the badly diseased animals, as they usually die or become badly stunted.

RABIES, HYDROPHOBIA.—Rabies is an infectious disease affecting the nervous system, that is transmitted by the bite of a rabid animal and the inoculation of the wound with the virus present in the saliva. It is commonly considered a disease of dogs, but because of the disposition of rabid dogs to bite other animals, rabies is common in domestic animals and man.

Rabies is widely distributed, being most prevalent in the temperate zone, and where the population is most dense. It has been excluded from Australia, Tasmania and New Zealand by a rigid inspection and quarantine of all imported dogs.

The specific cause of rabies is probably a protozoan parasite (the Negri bodies present in nerve-cells, Fig. 105). The germ spreads from the wounds through the nerves and central nervous system. The disease-producing organisms are present in great numbers in the nerve-tissue and saliva.

The period of incubation varies from a few days to several months. It is usually from ten to seventy days.

The symptoms differ in the different species. There are two forms of the disease: the *furious* and the *dumb*. The former is more common.

In the dog, the symptoms may be divided into three stages. The first, or *melancholy stage*, usually lasts from twelve to forty-eight hours. The animal's behavior is altered and it becomes sullen, irritable and nervous. Sometimes it is friendly and inclined to lick the hand of its master. An inclination to gnaw or swallow indigestible objects is sometimes noted. Frequently a certain part of the skin is rubbed or licked.

The second, or *furious stage*, may last several days. Violent nervous or rabid symptoms are manifested, and the dog may leave home and travel long distances. The animal usually shows a strong inclination to bite. It may move about snapping at imaginary objects in its delirium, and may bite any person or animal with which it comes in contact. The bark is peculiar, the appetite is lost and the animal becomes weak and emaciated.

In the third, or *paralytic stage*, the dog may present an emaciated, dirty, ragged appearance. The lower jaw may drop, the tongue hangs from the lips and the eyes appear sunken and glassy. Paralysis of the hind parts may be present.

In the dumb form, the paralytic symptoms predominate and the disease pursues a short course. Rabies terminates in death in from four to ten days.

Furious rabies is more common in the *horse*. The animal is very nervous, restless and alert. It may attack other animals in a vicious manner, kicking and biting them. The animal does not seem to care to eat or drink, and usually shows violent nervous symptoms, such as biting the manger, rearing and kicking when confined in the stable.

Cattle butt with the horns and show a tendency to lick other animals. They bellow more than common and the sexual desire is increased. Paralytic symptoms are manifested early in the disease, and the animal may fall when moving about. They soon present a gaunt, emaciated appearance.

In dogs the diagnosis is confirmed by a *microscopical examination* of the vagus ganglia and that portion of the brain known as Amnion's horn, and the finding of Negri bodies in the nerve-cells. In case a person is bitten by a dog, the animal should be confined until the disease is well advanced and killed or allowed to die. The head should then be removed and forwarded to the State laboratory, or wherever such examinations are made.

The treatment is preventive. Wherever an outbreak of rabies occurs all dogs should be confined on the owner's premises or muzzled. All dogs running at large without muzzles should be promptly killed. A heavy tax on dogs, and the killing of all dogs not wearing a license tag, would prevent the heavy financial loss resulting from rabies, and the ravages of wandering dogs in the United States. In countries where the muzzling of dogs is enforced during the entire year, rabies is a rare disease.

FOOT-AND-MOUTH DISEASE.—This is a highly contagious and infectious disease of cattle, sheep, goats and swine. It is characterized by the eruption of vesicles on the mucous membrane lining the mouth, the lips, between and above the claws and in the region of the udder and perineum. Man may contract the disease by caring for sick animals; or by drinking raw milk from a sick cow. Babies are most susceptible to infection from milk.

Foot-and-mouth disease was introduced into eastern Europe from the steppes of Prussia and Asia near the end of the eighteenth century. It was introduced into England about 1839, and in 1870 into Canada through the importation of cattle from England. From Canada the disease spread to the United States. Very few animals were infected during the 1870 outbreak, and the disease was quickly stamped out in both countries.

Europe has been unable to eradicate foot-and-mouth disease. The different outbreaks that occur from time to time cause enormous financial loss. In the United States outbreaks of the disease have occurred in the following years: 1870, 1884, 1902-'03, 1908 and 1914-'15. In the first two outbreaks very few cattle contracted the disease, and the infection was quickly stamped out. The third and fourth outbreaks were more extensive, and it was necessary to slaughter several thousand cattle and hogs in order to eradicate the disease. The first four outbreaks occurred in the eastern States, and the disease was prevented from spreading to the principal live-stock centers of the country, and the leading stock-raising States by slaughtering the diseased and exposed animals and by county and State quarantines. Early in the 1914-'15 outbreak, the disease spread to the Chicago Stock Yards, and from there, through shipments of cattle, to the principal live-stock sections of the country. The financial loss resulting from this outbreak has amounted to several million dollars. The Federal and State authorities have always been successful in stamping out the disease in the United States.

The specific cause of foot-and-mouth disease is a filterable virus that is present in the serum from the vesicles, the saliva, milk, and various body secretions and excretions from the sick animal. In the early stage of the disease it is present in the blood. None of the many investigators have been able to discover the microorganism that produces the disease.

Two of the outbreaks of foot-and-mouth disease in the United States originated from an infected vaccine used for the inoculation of vaccine heifers. The origin of the 1914-'15 outbreak has not been discovered. When introduced into a country, the disease spreads rapidly, through the movement of live-stock affected by the disease. Animals recently recovered may infect other animals. Dogs, birds, people, vehicles, milk, roughage, grains and other material from an infected farm may spread the disease.

The period of incubation is short. Symptoms of disease may be manifested in from one to six days following exposure.

The first symptoms are fever, dulness, trembling and loss of appetite. This is followed by vesicles or blisters forming on the mucous membrane of the mouth, lips, between and above the claws and the region of the udder. The inflammation of the mouth and feet may be very painful. Long strings of saliva may dribble from the mouth and collect about the lips. A smacking or "clucking" sound is produced when the animal moves its jaws and lips. The severe pain resulting from the inflammation of the mouth and feet, and the difficulty in moving about and eating and drinking, cause the animal to lose flesh and become emaciated. Milk cows may go dry.

The death-rate is not heavy. Some writers place it as low as two or three per cent. Because of the erosions and sloughing of the tissues of the mouth, feet and udder it becomes necessary to kill many of the animals. Young animals frequently die of inflammation of the digestive tract. The immunity conferred by an attack of the disease is not permanent.

The most economical measures of *prevention and control* are to buy and slaughter all diseased and exposed animals, bury the carcasses in quicklime, disinfect the premises (Figs. 107, 108 and 109) and enforce a district, county and State quarantine, until after the infection has died out. This statement may not hold true of methods of control in countries where foot-and-mouth disease is widely distributed.

TETANUS. LOCKJAW.—This is an acute infectious disease that is characterized by spasmodic contractions of voluntary muscles. The specific germ remains at the point of infection, and produces toxins that cause tetanic contractions of the muscles. It commonly affects horses, mules, cattle, sheep and swine. The disease is most common in warm, temperate climates.

The *specific cause* is a pin-shaped germ, the *Bacillus tetani* that is present in the soil, especially those that are rich and well manured. The germ enters the body by way of a wound, especially punctured wounds. Infection may take place through some wound in the mucous membrane lining the mouth, or other parts of the digestive tract. Infection may follow a surgical operation, such as castration. In any case, the germ requires an absence of air (oxygen) for its development.

The period of incubation varies from one to two weeks, the length of time depending on the nearness of the wound to a large nerve trunk or brain.

The first *symptom* observed is a stiffness of the muscles, especially those nearest the point of inoculation or wound. The muscles of the head, neck, back and loins are often affected first, and when pressed upon with the fingers feel hard and rigid. The disease rapidly extends, producing spasms of other muscles of the body. In breathing, the ribs show less movement than normal, the head is held in one position and higher than usual, the ears are stiff or pricked, the nostrils dilated, the lips rigid or drawn back and the eyes retracted, causing the "third eyelid" to protrude over a portion of the eye. In most cases the muscles of mastication and swallowing are affected. The animal may be unable to open its mouth and swallows with difficulty. When standing, the limbs are spread out so as to increase the base of support, and in acute cases about to terminate fatally, the pulse is quick and small and the respiration shallow, rapid and labored. The animal sweats profusely, falls down and struggles violently, but remains conscious to the end.

In the *subacute form* the symptoms are mild, and the animal may be able to move about, eat and drink without very great effort.

Treatment is largely preventive. All wounds should be carefully disinfected. This is especially advisable in punctured wounds of the foot. In communities, or on premises where tetanus is a common disease, animals that have punctured or open wounds should be given a protective dose of tetanus antitoxin.

The curative treatment is largely good care. If a wound is present, it should be thoroughly disinfected. The animal may be supported by placing it in a sling. A comfortable box-stall, where the animal is not annoyed by noises or worried by other animals, is to be preferred. A fresh pail of water should be given the animal several times daily.

The course of the disease varies. Death may occur within a few days, or the disease may last two or three weeks. Animals that recover from tetanus may show symptoms of the disease for several weeks. The death-rate is highest in hot climates and during the summer months.

If the animal can eat, it is not advisable to feed a heavy ration of roughage or grain. A very light diet of soft food, such as chops and bran-mash, prevents constipation and encourages recovery. Drugs that have a relaxing effect on the muscles may be given. Tetanus antitoxin may be given in large doses.

QUESTIONS

1. What is septicaemia and pyaemia?

2. What is haemorrhagic septicaemia? Give methods of spreading and controlling this disease.

3. Give the cause of anthrax and symptoms.

4. What control measures are recommended in anthrax?

5. What is ulcerative sore mouth? Give the treatment.

6. Describe the symptoms occurring in rabies, and state the control measures recommended.

7. Name the species of animals affected by foot-and-mouth disease, and the countries where the disease is prevalent.

8. Give the methods of distribution and control of foot-and-mouth disease.

9. What is the specific cause and method of infection in tetanus? Give the treatment.

CHAPTER XXIX
INFECTIOUS DISEASES OF THE HORSE

STRANGLES. DISTEMPER.—This is an acute infectious disease associated with a catarrhal condition of the air-passages and suppuration of the lymphatic glands in the region of the throat. Colts are the most susceptible to the disease. One attack renders the animal immune against a second attack of the disease, but the immunity is not always permanent.

The specific cause, Streptococcus equi was discovered by Schutz in 1888. Strangles is commonly spread by exposing susceptible animals to diseased animals, either by direct contact, or by exposing them to the infection in the stable and allowing them to drink or eat food from watering and feeding troughs on premises where the disease exists. The predisposing causes are cold and sudden changes in the weather. For this reason the disease is most prevalent during the late winter and early spring.

The period of incubation varies, usually from four to eight days.

The symptoms at the beginning of the attack are a feverish condition and partial loss of appetite. The visible mucous membranes are red and dry. This is followed by watery nasal secretions that become heavy and purulent within a few days. The inflammation may extend to the larynx and pharynx.

The glands in the region of the jaw become hot, swollen and painful, and the animal may be unable to eat or drink. The swelling and inflammation of the throat, and the heavy, pus-like secretions that accumulate in the nasal cavities, cause difficult respirations. After a few days the abscesses usually break, and the symptoms are less severe. If the abscesses break on the inside of the throat, the discharge from the nostrils is increased.

The disease may be accompanied by an eruption of nodules, or vesicles on the skin, or nasal mucous membrane.

In severe and chronic cases multiple abscesses may form. This complication is indicated by emaciation and weakness. Such cases usually terminate in death. Severe inflammation and swelling in the region of the throat may terminate in strangulation and death. The death-rate is from one to three per cent.

The preventive treatment consists in using all possible precautions to prevent the exposure of susceptible animals and practising the immunization of exposed animals. The curative treatment is principally careful nursing. Rest, a comfortable stall, nourishing feed and good care constitute the necessary treatment for the average case of distemper. When the abscesses become mature, they should be opened and washed with a disinfectant. Steaming the animal several times daily relieves difficult breathing and the irritated condition of the mucous membranes. In case the abscesses do not form promptly and the throat is badly swollen, a blistering ointment or liniment may be applied. Bitter and saline tonics, the same as recommended in the treatment of indigestion, may be given with the feed.

INFLUENZA (CATARRHAL OR SHIPPING FEVER).—This is a well-known acute infectious disease of solipeds. It is characterized by depression, high body temperature and catarrhal inflammation of the respiratory and other mucous membranes.

Several epidemics of influenza have occurred in the United States. The most serious epidemic occurred in the latter part of the '70's, and the last one in 1900-'01. Influenza is present in the principal horse centers in a somewhat attenuated form.

The specific cause of the disease has never been determined. The virus is present in the expired air, nasal secretions and excreta. Close proximity to a diseased animal is not necessary in order to contract the disease. Stables may harbor the infection, and it may be distributed by such disease carriers as blankets, harness, clothing of the attendant and dust.

The predisposing causes are cold, exposure and changes in climate. When the disease appears in a country, it is first present in the large cities, and from there it is scattered to the outlying districts. The *period of incubation* is usually from four to seven days.

The early symptoms of the disease are a high fever, marked depression and partial or entire loss of appetite. The horse usually stands in the stall with the head down and appears sleepy. The visible and respiratory mucous membranes are inflamed, the respirations are quickened and the animal may cough. The eyes are frequently affected, the lids and cornea showing more or less inflammation. The digestive tract may be affected. At the beginning, colicky pains may be present and later constipation and diarrhoea. Symptoms of a serious nervous disturbance are sometimes manifested.

The limbs usually become swollen or filled. This disappears as the animal begins to improve. Pregnant mares may abort. The death-rate is low.

The treatment required for the sick animals is largely rest, a light diet and a comfortable, clean, well-ventilated stall, free from draughts. Windows in the stall should be darkened. If the stable is cold, the body of the animal should be covered with a blanket and the limbs bandaged. Two ounces of alcohol and one drachm of quinine may be given three or four times daily. Small doses of raw linseed oil may be given if necessary.

Horses that are exposed to cold, wet weather or worked after becoming sick, frequently suffer from pneumonia, pericarditis, gastro-enteritis and other diseases. Such complications should be given prompt treatment.

It is very advisable to give a protective serum to horses that are shipped or transported long distances, and exposed to the disease in sale or transfer stables.

GLANDERS, FARCY.—This is a contagious and infectious disease of solipeds that is characterized by the formation of nodules and ulcers on the skin, nasal mucous membrane and lungs.

Although glanders is one of the oldest of animal diseases, it was not until 1868 that its contagious character was demonstrated. The disease is widely distributed. It became more prevalent in the United States after the Civil War. The vigorous control measures practised by the State and Federal health officers have greatly decreased the percentage of animals affected with glanders. At the present time the disease is more often met with in the large cities than in the agricultural sections of the country.

The specific cause of glanders is the Bacillus mallei . This microorganism was discovered in 1882. It is present in the discharges from the nasal mucous membrane and the ulcers. These discharges may become deposited upon the feed troughs, mangers, stalls, harness, buckets, watering troughs, drinking fountains and attendants' hands and clothing. Healthy horses living in the same stable with the glandered animals may escape infection for months. It is usually the diseased animal's mate, or the one standing in an adjoining stall, that is first affected. Catarrhal diseases predispose animals to glanders, as the normal resistance of the mucous membranes is thereby reduced. The most common routes by which the germ enters the body are by way of the digestive and respiratory tracts. It may enter the body through the uninjured mucous membranes of the respiratory tract and genital organs, or through wounds of the skin.

The period of incubation may be from a few to many days.

The symptoms may be *acute* or *chronic* in nature. The *acute form* pursues a rapid course. It is frequently seen in mules and asses, and it may develop from the subacute or chronic form in horses. When the disease is acute, the animal has a fever, is stupid, does not eat, and may have a diarrhoea. In this form the lymphatic glands suppurate, the animal loses flesh rapidly and dies in from one to two weeks.

The *chronic form* is the most common. It develops slowly and lasts for years. The early symptoms of the disease (chilling and fever) usually escape notice. The first visible symptom is a nasal discharge of a dirty white color from one or both nostrils. This is usually scanty at first, and intermittent, but later becomes quite abundant. The discharge is very sticky, and adheres to the hair and skin. The most frequent seat of the disease is in the respiratory organs, lymph glands and skin. Nodules and ulcers appear on the nasal mucous membrane but they may be so high up as to escape notice. The ulcers are very characteristic of the disease. They are angry looking, with ragged, raised margins, and when they heal leave a puckered scar. The submaxillary glands may be enlarged, and at first more or less hard and painful, but later they become nodular and adhere to the jaw or skin. Nodules and ulcers may form on the skin over the inferior wall of the abdomen and the inside of the hind limbs and are known as "farcy buds." Lymphatic vessels near these buds become swollen and hard. The animal loses flesh rapidly, does not withstand hard work, and the limbs usually swell.

It is sometimes difficult to diagnose the disease. The ulcers on the nasal mucous membranes and elsewhere are very characteristic, and when present enable the examiner to form a diagnosis. In cases of doubt, a bacteriological examination of the nasal discharge may be made, or we may resort to one or several of the various diagnostic tests. The Mallein test is quite commonly used. The sterilized products of a culture of the *B. mallei* are injected beneath the skin of the suspected animal. This causes a rise in body temperature and a hot, characteristic swelling at the point of injection in glandered animals.

Treatment is not recommended at the present time. Nearly all of the States have laws which aim to stamp out the disease wherever found by killing all affected animals, and thoroughly disinfecting the stables, harness and everything which has been near the animal. Diseased animals should be carefully isolated until slaughtered, and all animals exposed to them should be subsequently tested for glanders.

CONTAGIOUS PLEUROPNEUMONIA (STABLE PNEUMONIA).— This is an infectious disease of solipeds that usually results in a fatal inflammation of the lungs and pleural membrane.

Many writers have described this disease as associated with influenza, but it is frequently seen as a separate disease, usually involving only the lungs and pleurae. It is prevalent in several parts of the United States, more particularly the horse centers or large markets, where it appears in the form of epidemics. In several of these localities it is known as western or stable fever.

The specific cause is not definitely known. The *Streptococcus pyogenes equi* is very commonly present. This germ grows in the diseased tissues. The disease is spread by direct or indirect contact, as when well or susceptible animals are placed in the same stable with an animal affected with the disease, or in stalls which have recently held diseased animals.

The period of incubation is from four to ten days following exposure.

The symptoms are those commonly seen at the beginning of an attack of simple pneumonia and pleurisy. They consist of chills, high fever, cough, depression, difficult and labored breathing and loss of appetite. The disease usually runs a course of from one to three weeks. The death-rate is thirty per cent or more.

The treatment is mainly preventive. Stables where horses having pleuropneumonia have been kept should be cleaned and disinfected by spraying the floors, stalls and walls with a four per cent water solution of a cresol disinfectant. It is advisable to subject all newly-purchased animals to a short quarantine period before allowing them to mix with the other animals in the stable. Exposed animals may be given a protective serum.

The curative treatment is the same as recommended for the treatment of simple pneumonia and pleurisy.

QUESTIONS

1. What is the specific cause of distemper? Give the symptoms and treatment.

2. What are the different methods of spreading influenza? Give the symptoms and treatment.

3. Give the cause and methods of controlling glanders.

4. Give the cause and treatment of contagious pleuropneumonia.

CHAPTER XXX
INFECTIOUS DISEASES OF CATTLE

ACTINOMYCOSIS, "LUMPY JAW."—This is an infectious disease that is characterized by the formation of tumors and abscesses and the destruction of the infected tissues. The disease is common in cattle and usually affects the bones and soft parts of the head. In the United States, where the disease is known as "lumpy jaw" the jawbone is commonly affected. In European countries the disease frequently involves the tongue, and the term "wooden tongue" is applied to it. The disease may affect regions of the body other than the head. Actinomycosis of the lungs sometimes occurs. Swine and horses may be affected by this disease.

The specific cause of actinomycosis is commonly known as the ray fungus. This fungus grows on certain plants, and the animal usually contracts the disease by eating plants or roughage that have the fungus on them. Grasses having awns that are capable of wounding the mucous membrane of the mouth and penetrating the gums are most apt to produce the disease. Young cattle that are replacing and erupting their teeth are most prone to "lumpy jaw." Conditions that favor bruising of the jaw and external wounds favor the development of actinomycosis.

The fungus grows in the tissues, causing an inflammatory reaction and destruction of the tissue. The ray fungus can be seen in the diseased tissue or the pus as yellowish, spherical bodies about the size of a grain of sand. Each of these bodies is formed by a large number of club-like bodies arranged about a central mass of filaments.

The local symptoms are characteristic. The tumor may involve the soft tissues of the head. If the jawbone is affected the tumor feels hard and cannot be moved about. Sometimes it is soft and filled with pus. Tumors of long standing may possess uneven, nodular surfaces and fistulous openings. When the tongue is affected, it is swollen and painful, and prehension and mastication of the food may be impossible. When the pharynx is the seat of disease, breathing and swallowing are difficult and painful. Actinomycosis of the lungs may present the appearance of a chronic pulmonary affection.

If the disease involves the head and lungs, the animal may become unthrifty and emaciated. In doubtful cases a microscopic examination of a piece of the tumor, or some of the pus, may be necessary.

The treatment is surgical and medicinal. Small, external tumors may be removed by an operation. Sometimes an incision is made into the diseased tissue and a caustic preparation introduced.

The most desirable method of treatment is the administration of large doses of iodide of potassium in a drench, or in the drinking water. The dose is from one to three drachms daily for a period of seven to fourteen days. The size of the dose depends on the size of the animal and its susceptibility to iodism. An animal weighing 1000 pounds may be given two drachms. The treatment is kept up until the symptoms of iodism develop. The condition is indicated by a loss of appetite and a catarrhal discharge from the eyes and nostrils. When this occurs, the treatment should be stopped, and the animal drenched with one-half pound of Epsom salts, and the dose repeated after three or four days. After an interval of two weeks, the iodide of potassium treatment should be repeated if the growth of the tumor is not checked.

EMPHYSEMATOUS ANTHRAX, "BLACK LEG."—"Black leg" is an acute infectious disease of cattle that is characterized by lameness and superficial swellings in the region of the shoulder, quarter or neck. The swellings are hot and painful and usually contain gas.

The specific cause of "black leg" is a rod-shaped, spore-producing germ, the bacillus of emphysematous anthrax. This germ possesses great vitality, and may live indefinitely in the soil. It has been known to live for years in clay and undrained soils. Young animals that are in high condition are predisposed to the disease.

The germ enters the body through abrasions in the skin and mucous membrane of the mouth and intestines.

"Black leg" is a common disease of young cattle in all sections of the country where cattle-raising is engaged in extensively. Outbreaks of the disease are most prevalent in the early spring after the snow has melted, and in the late summer in localities where cattle graze over the dried-up ponds and swampy places in the pasture. The germs of black leg may be carried from a farm where the disease is prevalent to non-infected premises by surface water. The opening up of drainage ditches through stock-raising communities may be followed by outbreaks of the disease.

The symptoms of black leg develop quickly and may terminate fatally in a few hours. These are general dulness, stiffness, prostration and loss of appetite. Lameness is a prominent symptom. The animal may show a swelling in the regions of the shoulder and hindquarters or on other parts of the body. The swelling is very hot and painful at first, but if the animal lives for a time, it becomes less tender, crackles when pressed on and the skin may feel cold and leathery. Fever is a constant symptom. In the highly acute form of the disease nervous symptoms, such as convulsions and coma, occur.

The tissue changes in the region of the swelling are characteristic. An incision into the swelling shows a bloody, dark exudate and the surface of the muscular tissue is dark. Frothy, bloody liquid escapes from the mouth, nose and anus.

The preventive treatment consists in thoroughly draining pastures and yards where cattle run. This measure does not insure cattle against the disease. Cattle that die of "black leg" should be cremated. This should be done at the spot where the animal dies. If the carcass is moved or opened, the ground should be thoroughly wet with a four per cent water solution of a cresol disinfectant and covered with lime.

Vaccination of the exposed or susceptible animals should be practised. On farms where the disease exists it may be necessary to vaccinate the young animals (less than two years of age) once or twice every year in order to prevent the disease. Medicinal treatment is unsatisfactory.

TEXAS OR TICK FEVER.—Tick fever is an infectious disease of cattle. It is caused by an animal organism that is present in the blood, and is conveyed from the animal that is host for the tick fever parasite to the non-infected animal by a tick (Figs. 120 and 121).

Tick fever was introduced into the southern portion of the United States through importation of cattle by the Spaniards. Previous to the establishing of a definite quarantine line between the permanently infected and the non-infected sections, heavy losses among northern cattle resulted through driving and shipping southern cattle through the northern States. The specific cause and the part taken by the tick in its distribution were not discovered until 1889-'90. Smith recognized and discovered the specific cause of the disease, and Kilborn and Salmon proved by a series of experiments that the cattle tick was responsible for the transmission of the disease from animal to animal.

The specific cause of tick fever is a protozoan parasite, *Piroplasma bigeminum* . It is present in the blood of cattle that are affected with this disease. The natural method of entrance into the body is through the bite of the cattle tick. The disease may be transmitted by inoculating blood containing the parasite into a susceptible animal.

There are two forms of the disease, the *acute* and *chronic*.

The symptoms of the acute form of the disease are a high fever, depression, loss of appetite, diarrhoea, dark or bloody urine, staggering gait and delirium. Death may occur within a few days from the time the first symptoms are manifested.

The symptoms of the chronic form of the disease resemble the acute form, but are more mild. The animal is unthrifty and loses flesh rapidly. The bloodless condition of the body is manifested by the pale, visible mucous membrane. Death seldom occurs.

The most characteristic *diseased changes* found on post-mortem examination occur in the liver and spleen. The liver is enlarged, and a yellowish, mahogany-brown color. The gallbladder is filled with a very thick bile. The spleen may be several times the normal size and dark colored. When it is cut into, the pulpy tissue may resemble thick, dark blood. The kidneys are pale and the bladder may contain dark or reddish-colored urine.

In the northern States and outside of the quarantined area, the direct or indirect exposure of the affected cattle to southern cattle, and the presence of the cattle tick, *Margarophus annulatus*, are sufficient evidence to confirm the diagnosis of this disease.

The prevention and control depend on destruction of the cattle tick. In the early history of the disease, shipping and driving of southern cattle into and through the northern States caused outbreaks of tick fever and heavy losses among northern cattle. This finally resulted in the locating of the infected district, and the establishment of the Texas-fever quarantine line in 1891 by Dr. D. E. Salmon.

Previous to this time Kilborne and Salmon proved that the cattle tick was essential to the spread and production of the disease. A further study of the life history of the tick resulted in the discovery that it could not mature unless it became a parasite of horses, mules, or cattle. This has led to the eradication of the tick in certain sections of the South, by not allowing cattle access to a pasture or lot for a certain period, and by freeing the animals from ticks by hand-picking, dipping and smearing.

The immunization of cattle that are shipped into an infected district for breeding purposes is often practised. Immunity is obtained by introducing the P. bigeminun into the blood, either by placing a few virulent young ticks upon the animal, or by repeated inoculation with a very small quantity of virulent blood.

QUESTIONS

1. Give the cause and treatment of actinomycosis.

2. Give the cause and treatment of emphysematous anthrax.

3. Give the cause of tick fever; distribution of the disease and methods of control.

CHAPTER XXXI
INFECTIOUS DISEASES OF POULTRY

FOWL CHOLERA.—This is a highly infectious disease of all species of poultry, that is characterized by weakness, depression and yellowish colored excrement.

The *specific cause* of fowl cholera is the *Bacillus avisepticus* . This microorganism is transmitted to the healthy birds by the feed, or water becoming contaminated with the discharges from the diseased birds. According to Salmon, the period of incubating varies from four to twenty days.

The early symptoms are a falling off in appetite, high fever, dulness, diarrhoea and weakness. The affected bird becomes drowsy, the head is drawn toward the body, and it may remain asleep for long periods at a time. Salmon states that the general outline of the sick bird becomes spherical or ball-shaped.

The disease is usually highly fatal. In the acute form the larger portion of the flock may die off within a week. In the subacute and chronic forms, the birds become greatly emaciated, and a few die off weekly through a period of a month or longer.

The tissue changes occurring in the disease are inflammation of all or a few of the internal organs. Ward states that the most characteristic lesion of fowl cholera is the severe inflammation of that portion of the small intestine nearest to the gizzard. Small hemorrhagic spots may be found on the heart and other organs.

The treatment is both preventive and curative. The preventive treatment consists in quarantining newly purchased birds until we are satisfied that they are free from disease. The occasional disinfection of the poultry houses and runs is highly important. Cleaning the poultry house by removing the floor, roosts, or any part of the house for the purpose of removing all filth, and spraying the interior with a three per cent water solution of a cresol disinfectant, should be practised. Lime should be scattered over the runs, or the yards immediately about the house. The above preventive measures form an important part of the care and management of the flock. The carcasses of the dead birds should be burned. It is advisable to kill all birds that are fatally sick.

All of the flock should be given antiseptics with the feed and water. Four ounces of a water solution of copper sulfate, made by dissolving one-quarter pound of this drug in one gallon of hot water, may be added to each gallon of drinking water. Frequent disinfection of the drinking fountains, feeding places and houses should be practised.

DISEASES RESEMBLING FOWL CHOLERA.—There are a few diseases, such as septicaemia, limber neck and infectious enteritis, that are sometimes mistaken for fowl cholera. These diseases are caused by different microorganisms that may be found in the digestive tract and air-passages of healthy birds, insanitary conditions and decomposed feed, especially meat. It seems that under certain conditions, such as insanitary quarters and birds that are low in constitutional vigor and weakened from other causes, certain germs may become disease-producers. The death rate from mixed infections is very heavy in poultry.

The symptoms vary in the different cases. The disease may be highly acute, as in limber neck, or chronic, extending over a period of a week or more. Diarrhoea is not a prominent symptom in the majority of cases.

The post-mortem lesions vary from a hemorrhagic to a chronic inflammation of the different body organs and serous membranes.

The treatment is preventive. A frequent cleaning and disinfecting of the poultry house and surroundings, avoiding the feeding of spoiled feed, or allowing the drinking fountains and feeding places to become filthy, are effective preventive measures. Sick birds should be either isolated and quarantined, or destroyed. Antiseptics may be given with the feed and drinking water.

AVIAN DIPHTHERIA (ROUP).—This infectious disease of poultry is especially common in chickens. It is characterized by a catarrhal and diphtheritic inflammation of the mucous membranes of the head.

The specific cause of roup has not been determined. The disease-producing germs are present in the discharges from the nostrils, eyes and mouth, and the body excretions of sick birds. Birds having a mild form of roup, or that have recently recovered from it, are common carriers of the disease. The disease is usually introduced into the flock by allowing birds exposed at poultry shows, or recently purchased breeding stock from an infected flock, to mix with the healthy birds.

The predisposing causes are very important factors in the development of roup. Cold, damp, draughty, poorly ventilated poultry houses cause the disease to spread rapidly and become highly acute.

The symptoms differ in character in the different outbreaks of the disease. Usually the first symptoms noticed are sneezing, dulness, diminished appetite and a watery discharge from the nostrils and eyes. Later the eyelids may become swollen and the nostrils plugged by the discharge from the inflamed membranes. If the mouth is examined at this time, an accumulation of mucus and patches of diphtheritic or false membranes are found. In the acute form of roup the false membranes and yellowish, cheesy-like material accumulate on the different mucous membranes, and interfere with vision, breathing and digestion. The affected bird becomes thin and weak. The death rate is very high in this form of the disease.

The preventive treatment consists in quarantining birds that have been purchased from other flocks, and that have been exhibited, for a period of three weeks. A careful examination of the mouth should be made. If a catarrhal discharge from the nostrils and false membranes is present, prompt treatment should be used. A sick bird should be held in quarantine for several weeks after it has recovered, and receive a thorough washing in a two per cent water solution of a cresol disinfectant before allowing it to mix with the healthy birds.

The medicinal treatment consists in removing the discharges from the nostrils and eyes with pledgets of absorbent cotton that are soaked with a four per cent water solution of boric acid. Among the common treatments mentioned are boric acid and calomel, equal parts by weight, blown into the nostrils and eyes with a powder blower. Water solutions of boric acid, potassium permanganate and hydrogen peroxide are recommended. Liquid preparations are applied with pledgets of cotton, oil cans, or atomizers.

Many recoveries can be obtained with careful treatment. It is usually most economical to kill the severely affected birds. Many poultrymen dispose of the entire flock as soon as the disease makes its appearance, and clean and disinfect the premises before restocking.

CHICKENPOX.—In some sections the disease appears in another form, known as *chickenpox* (contagious epithelioma), in which nodules form on the skin along the base of the comb and other parts of the head, or both forms may be met with in the same flock. The nodules should be treated with vaseline, or glycerine ointments containing two per cent of any of the common antiseptics or disinfectants.

ENTERO-HEPATITIS. "BLACKHEAD."—This is a very fatal disease of young turkeys. Grown turkeys and other fowls are not so susceptible to the disease. It is characterized by an inflammation of the liver and intestines, especially the caeca.

The specific cause is a protozoan microorganism, *Amoeba meleagridis*. Adult fowls and turkeys may act as carriers of the germ, and the young turkeys become infected at an early period.

The symptoms are diminished or lost appetite, dulness, drooped wings, diarrhoea, weakness and death. When the disease becomes well advanced, the head and comb become dark.

The course of the disease is from a few weeks to three months. Very few of the young turkeys survive.

The treatment is almost entirely preventive. The same precautionary measures for the prevention of the introduction of disease into the flock, recommended in other infectious diseases, should be practised. Turkeys that survive should be disposed of. As chickens may harbor the disease-producing germs, we should not attempt to raise turkeys in the same quarters with them. Eggs should be obtained from disease-free flocks. Wiping the eggs with a cloth wet with fifty per cent alcohol may be practised. The same recommendations regarding the cleaning and disinfecting of the quarters described in the treatment of fowl cholera should be practised.

If an outbreak of the disease occurs in the flock all of the sick birds should be killed, and their carcasses cremated. Moving the flock to fresh runs and the administration of intestinal antiseptics are the only effective lines of treatment.

AVIAN TUBERCULOSIS.—Tuberculosis of poultry is a serious disease in some countries. Poultry usually contract tuberculosis by contact with a tubercular bird, and not from other domestic animals and man.

The symptoms are of a general character, such as emaciation, weakness, wasting of muscles and lameness. Tubercular growths may appear on the surface of the body.

If we suspect the presence of the disease, it is advisable to kill one of the sick birds and make a careful examination. The finding of yellowish, white, cheesy nodules or masses in the liver, spleen, intestines and mesenteries is strong evidence of tuberculosis. A bacteriological examination of the tissues may be necessary in order to confirm the diagnosis.

The same *methods of treatment* as recommended in tuberculosis of other domestic animals may be used in eliminating the disease from the premises and flock. This consists in killing and cremating all birds showing visible symptoms, moving the apparently healthy portion of the flock to new quarters and wiping the eggs with alcohol. The old quarters should be cleaned, disinfected, and then allowed to stand empty for several months, when we should again spray with a disinfectant, and scatter lime over the runs. If the cleaning and disinfecting have been thorough, we may safely turn young or healthy birds into the old quarters. All possible precautions against carrying the infection to the healthy flock must be observed.

QUESTIONS

1. Give the cause and treatment for fowl cholera.

2. What diseases resemble fowl cholera? Give the treatment.

3. Give the symptoms and treatment for roup.

4. Give the treatment for "blackhead."

5. Give the treatment for Avian tuberculosis.

REFERENCE BOOKS

Pathology and Therapeutics of the Diseases of Domestic Animals, Vol. I-II, Hutyra and Marek.

Veterinary Medicine, Vol. I-V, Law.

General Therapeutics for Veterinarians, Frohner.

Prevention and Treatment of the Diseases of Domestic Animals, Winslow.

Age of the Domestic Animals, Huidekoper.

Veterinary Materia Medica and Therapeutics, Winslow.

Veterinary Anatomy, Sisson.

Chauveau's Comparative Anatomy of Domestic Animals.

Manual of Veterinary Physiology, Smith.

Annual Reports of Bureau of Animal Industry, from 1902 to 1911.

End of Project Gutenberg's Common Diseases of Farm Animals, by R. A. Craig